THE NATURE OF NATURE

THE NATURE OF NATURE

THE NATURE OF NATURE

The cover has a NASA photo from the Galileo, the solar system explorer, showing the earth, the teeming blue planet, with its barren moon, devoid of life.

The cell structure diagram on the back cover comes from http://www.webcalf.com

The dinosaurs and Möbius strip come from Wiki commons.

COLIN MAYNARD PRICE

The nature

of

Nature

The character of science and theology and other valuations

THE NATURE OF NATURE

CONTENTS

THE NATURE OF NATURE

The nature of Nature

THE NATURE OF NATURE

EVOLUTION AND NATURAL SELECTION

Biologists, almost without exception, carefully avoid confusing "evolution" with "natural selection".

This is not so in the popular mind where *evolution* is often the preferred term to *natural selection*. Curiously, the word "evolution" doesn't occur in Charles Darwin's *The origin of species* (1859); *w*hich to give his theory the full title is *The Origin of Species by means of Natural Selection or the preservation of favoured races in the struggle for life*. And it is the words *Natural Selection* that are the key to Darwin's ideas.

Darwin avoided the word *evolution* with good reason. It was already a popular and easily misunderstood word through Robert Chamber's *Vestiges of the Natural History of Creation* (1844) but even before that it had been used by Darwin's grandfather Erasmus to much later embarrassment. Then as now evolution is an ill-defined term, or more precisely a multi-variably defined term which means different things to different people.

Darwin was interested in the origin of species because along with many he no longer held the view of the "fixity of the species" - the view that a select number of species had been created at a universal beginning.

People had been collecting species for over a hundred years and continued to do so throughout the 19th century with the results carefully mounted in mahogany cabinets; species now were numbering thousands and thousands. Clearly, the emergence of many thousands of new species needed explanation.

Darwin (and even more so his followers) studiously avoided two possible explanations. One, that of Lamarck, that the "inheritance of acquired characteristics" (acquired during life that is) could gradually change one species into another. And the other, that acquisition of novel new features like the mammalian eye could be the result of an interposing Designer along the lines of Paley's intelligent Watchmaker.

Darwin noted with interest the techniques of husbandry and horticulture in modifying cattle, sheep, dogs, pigeons, crops and plants through *human selection* and soon began to work out how a *natural selection* might operate.

Natural selection, Darwin supposed, is primarily brought about by a general pressure for survival caused by an over-production of off-spring and an over-demand on scarce food resources. This he took from Malthus' gloomy analysis on populations and food production.

Secondly, Darwin supposed that these over-abundant offspring while breeding true also varied slightly in a random way. It was this "random occurrence of variation in offspring" which, with all other things being equal, gave some off-spring an advantage. Those with any advantage were better fitted for survival other things being equal. Of course it could be, as in human life, that even those with all the advantages might not survive for other accidental reasons. But in the long term those best fitted to survive were more likely to survive.

Darwin's thought is in terms of averages over long periods of time over many generations of different struggles both within species and between species. So on average under threats of annihilation, where most will die young anyway those that survive *to reproduce* (that's important) will tend to do so if they have advantageous features over their competitors. And, this is the point, they will also tend to pass on these advantages to their off-spring who in turn will reproduce or not as the endless struggle persists.

Over time these selective changes will cause diversion in the line of generation as species adapt to their changing environment and eventually to separate species (that is, pairs that can no longer interbreed).

Such a rationale seemed to explain well the experimental and observational work Darwin carried out from HMS Beagle in the Galapagos Islands. There, he witnessed different species of finches and turtles, geographically separated and widely distributed among the different islands in the archipelago. Under isolation species begin to diverge, but evolutionists would want to explain a lot more and Darwin himself wanted to see his theory working out in the fossil record and not just in living populations.

They (the evolutionary biologists) would like to say (and do say if not in so many words) that the beaks of finches evolved as adaptations to new environments and that birds themselves evolved from the dinosaurs and that the tetrapods themselves evolved from the fishes and so on down to eukaryotic cells which evolved from cells without nuclei, called prokaryotic cells.

However crudely or sophisticatedly it is put the difference as to what evolutionists say and what Darwin says as the essence of natural selection is this. In any event A and some subsequent event B (which might be a feature of beak shape on one finch but an altered feature on another finch; or some order of species like the dinosaurs with another, event B, the aves; or some species like the velvet worms which lie betwixt the phyla of nematodes and arthropods) then the evolutionary biologist says there is some relationship between event B and event A. It might not be causal (except perhaps as a long genetic "fuse") but there is a linkage. Event B evolved or was an adaptation from event A or event A gave rise to this or that event B.

However under the strict terms of natural selection, which is brought about by random and varied characteristics it is absolutely clear that there is no relationship between event A and event B. Darwin was quite clear that there was "no law of necessary development" at work. In a word, no evolution. There is no relationship between events A and B. That was the whole point of natural selection. It was natural. It didn't require any additional *élan vital* by way of explanation.

Yet it is vital for the theory of evolution (as it developed) that whatever might be said in a retrospective perspective, for instance, that hominids look like they evolved from the primates or whatever, the tiny changes that together make up each big leap are each independent, unrelated, random or mutational. Such mutations might be associated with transcription errors, but each and every transcription error is non-predictable. Each occurs randomly.

This is why unless care be taken evolutionary notions are at odds with the tenets of natural selection.

But it begs the question; how can it be that the big leaps which are steps in evolution are made up of tiny leaps which are entirely random and are typically transcription errors in the cell chemistry?

If one didn't know better one might be forgiven for thinking that primarily, by some means or other, not presently known, first came the big leaps of change, triggered by one knows not and then, additionally the small, random, modifications selected for particular environments. Not the other way round.

THE NATURE OF LIFE

Natural selection after Charles Darwin was modified by two important developments into what is sometimes called neo-Darwinism.

First, although Darwin knew nothing of the Mendel's laws of hereditary which only became widely known at the beginning of the 20th century he knew enough of breeding techniques to suppose that off-spring bred true to their parents but with a varied or random spread.

This role of the random in natural development is the key point in understanding natural selection. The chief architect in biological change is not a mechanical or predictive process but the chance variations within an over-production of a given population which because of survival pressure allows only the "fittest" to go on to reproduce the next generation. This was Darwin's understanding which Mendel's laws if anything gave additional support.

The random element in inheritance in the classic forms of Mendelism is simply the exchange of gene characteristics paired up from the two parents. For any one characteristic the gene would be either be dominant or recessive. What the off-spring inherits is a 50%/50% or, at worse a 25% possibility in the case of two recessive genes.

In terms of evolution the variation is within *both* the parents *and* in the gene pool of that breeding population. So there is also randomness around as individuals select mates. But this variation is not going to change much. A more important source of random change was in the early 20th century thought to be mutations. Mutations, before Crick and Watson and the discovery of the structure of DNA (1953) were thought largely due to cosmic rays and radio-activity. Both of these effects were random and caused chromosome damage (which came out in subsequent generations). This was pursued through much study of the unfortunate and much put-upon, fruit fly, *Drosophila melanogaster*

But post-Crick and Watson, in the modern interpretation, Darwin's random "variations" so essential to natural selection though still termed *mutations* are now almost exclusively assumed to be transcription errors in genetic copying.

Replication and transcription errors are not very numerous, there being correction procedures and redundancies (which make actual errors non-critical).

Peter Atkins writes.

"The copying proceeds at almost machine-gun rates: vertebrate RNA polymerase duplicates about thirty bases a second, and takes about seven hours to replicate a cell's full complement of DNA. About one base in a million is copied wrongly, but proof-reading enzymes keep a watchful eye and most errors are corrected leaving, about one error in about 10 billion bases." (1).

But how, we might ask, can evolutionary changes be brought about from random errors in the linear, one dimensional strand of DNA, which is to be ultimately expressed two spatial dimensions higher up as a complex, three dimensional protein, such as haemoglobin, which has a molecular weight of around 64,500 and is a three dimensional jig-saw puzzle of immense proportions? According to Max Perutz who was the first understand its mechanico-chemical structure, its four giant molecules go "click, click, click" (2) as it transports oxygen from the lungs to round the body. In human bone marrow red blood cells are manufactured at the rate of 2 million/sec and each red blood cell contains about 280 million haemoglobin molecules. Quite a feat of manufacturing to have been produced by random mutations along a linear strand of DNA!

Mutations through errors in copying can hardly be regarded as being responsible for the major changes in the evolutionary history of life on Earth. There is just not enough bang to produce such staggering results. There must surely be other factors at work.

This brings us to the second advantage we have over Darwin – our knowledge of the role of DNA as the essential common factor to all living things. In a cosmic sense the Earth has but one living bioform, spread out in many varieties and species admittedly, but not compartmentalised and isolated each from each other, as once understood by Darwin, the Bible and indeed the whole world.

The division of knowledge, which turns upon Crick and Watson's work starting in 1953, is no less monumental than the flat Earth paradigm change or the Copernican revolution.

Darwin's question on "what is the origin of species?" is replaced with "what is the origin and nature of DNA?" and "how come we are all related together – not only the races of mankind but everything that is alive or has been alive?"

No longer is the question about the evolution or natural selection of phenotypes (the individual changes in the *outward* appearance of creatures or plants) but about the unfolding of the cosmic genotype (the varieties of DNA *within* the whole mass of living forms).

This change makes a big difference. It changes the nature of the task set by Charles Darwin.

References

1. Peter Atkins *Galileo's Finger* (2003) p71

2. Max Perutz *Molecular biology at Cambridge* in *Cambridge Minds* ed. Richard Mason (1994) p203

THE QUESTION OF THE ORIGIN AND PURPOSE OF LIFE

Throughout most of human history a firm distinction has been made between the mineral and the living worlds (and sometimes called vitalism). In the early days of chemistry the distinction was between organic and inorganic chemical reactions.

Then, in 1828 Friedrich Wohler synthesised urea and the distinction became less clear, though in practice most organic chemicals still couldn't be manufactured without the aid of a living system. Examples of living systems would include those that made use of brewer's and baker's yeasts in the manufacture of beer and bread respectively.

In 1922 the Russian Aleksandr I Oparin had the theory that life might have formed in a chemical soup of simpler compounds. Darwin himself had mentioned something similar as a possibility.

In 1952 Stanley Miller and Harold Urey set up experiments to test Oparin's theory in a reducing atmosphere (ours is now oxidising) in the presence of energy in the form of lightning organic chemicals might be produced outside living systems and point to how life might have evolved. In an atmosphere of ammonia, methane, water vapour and other compounds they attempted to simulate the "primitive soup" of early Earth.

Since then all students now take it for granted that there is no real distinction between the organic and mineral worlds except in complexity. But it is the immense complexity of living systems that can't be ignored and ultimately forms a distinctive qualitative difference in the material existence on this planet.

The undergraduate text books on Biochemistry (1) tend have an initial section on the *chemical* evolution of macromolecules; putting this as a preliminary to *biochemical* evolution and prior to *biological* evolution and they always mention these experiments of Miller and Urey as a possibility for the origin of life, from basic building blocks. The building blocks of life are the monomers, like the nucleotides, the sugars and the amino acids. These are the building blocks for the macro or biomolecules, the deoxyribonucleic acids, polysaccharides and proteins.

The long list of synthesised products from the Miller-Urey experiments might look convincing, as including some of the building blocks, but clearly no one anywhere can either believe or think this is the actual way life evolved, for clearly if they did every "biochem lab" the whole world over would have been generating self-replicating macromolecules wholesale during the last fifty years! Which is patently not so. And which is also misleading, for the impression is given that such a process while not quite right, is along the right lines, whereas the fact is it's not even remotely on the right lines. It's a cul-de-sac masquerading as a short cut. The answer to both questions, the origin of life and the nature of life, is not going to be trivial. It is often hoped that given problems have simple and concise solutions, but it is also important to recognise the size and scope of the problem and not be self-deceived.

Incidentally these, two questions, the origin and nature of life, were not the problems to which Charles Darwin addressed himself (see Part 1) though the theory of natural selection and with it evolution has often been assumed, falsely, to also apply.

These two, questions, the origin and nature of life, are of fundamental importance and ought not to be trivialised or dismissed lightly. They stand proud over any other questions that can be asked. Philosophically we might ask what is the nature of reality and with it seek a knowledge of the world in science and technology and all the creative arts; theologically we can ask questions concerning the existence of God and how he is revealed; of ourselves we can ask what is the best way of living or as the Greeks put it "what is the good life?" but to ask our twin questions on the nature and origin of life is to dig deeper than any of these questions and to be confronted with the whole mystery of that indubitable fact that some things are living and some are not living. And it's not at all a religious question. (It's not entirely non-religious either.)

That it's a real problem is as undeniable as a sock on the jaw (as Wittgenstein would have put it). Indeed it is because it is so obvious and doesn't fit neatly or exclusively into any category, religious or scientific that it feels so primordially fundamental to every aspect of mankind's conscious existence, scientific, religious and ethical.

It is for these reasons that I now consider these questions seriously in Part 4.

References

1. Two typical text books I have on my shelves are *Principles of Biochemistry* by Lehninger, Nelson and Cox (2nd edition New York, 1993) and *Biochemistry* by Geoffrey Zubay (3rd edition, USA, 1993) – both originally multi-volume but cased as one.

THREE ANOMALIES OF
LIVING THINGS

The first of three anomalies is the oddity of the close commonality of all living things. This oddity became plain after Crick and Watson uncovered the significance of the macro-molecule, DNA in 1953.

All previously ages, of course, recognised the difference between things that live and grow and reproduce and things which do not. But even Charles Darwin would have been surprised to learn just how intimate are the connections. The strangeness is just how close we are to each other. Comparative genomics indicates minute if not zero differences between our human races; better than 95% between us and the chimps, quite a lot in common with worms and so on. With something like 3 billion base pairs in the human genome, the newt has 20 billion base pairs. Chromosome counts are no more encouraging to human pride with chickens having a diploid chromosome number of 78 to our 46. On gene estimates, the protozoan causing the disease *trichomoniasis* has at 60,000 nearly three times as many genes as us. We are in the thick of it, but hardly the crown of creation. Indeed there seems no great hierarchy. Every living thing is dependent on other living things. All creatures, all living things democratically draw from common biochemical enzymes and reactions. It is almost as if there is on Earth just one living bioform spread out in varieties of exploration. Our Earth bioform is more simply called, Life.

The second anomaly is the sheer complexity of the living world. Not just the variety and fecundity of "the great chain of being" as Lovejoy (1) called it, but the complexity of biochemistry, compared to inorganic chemistry (and where even the names of the chemicals are considerably longer).

The cell, the basis of all living things (except viruses, but essential to them also) is small - the dot on the "i" might cover two hundred typical cells. Magnify a typical cell 2 billion times, to the size of an oil refinery, say, and there would be general panic, fear and amazement at the complexity of such an alien-looking industrial complex working at such a high level of efficiency and productivity. If we were to stand back and look dispassionately at the nature of Nature we would see it as both wondrous and very puzzling.

But one complication against clear vision is that the living world is so intimate with the non-living that any original relationship between the two has been lost. The planet has been "terraformed" by the Earth bioform, Life, but is it commensal, symbiotic or parasitic? Or is it just the case of a plain and simple conqueror of territory, of the Genghis Khan variety? And where does this thing begin and end now?

Compared to the other planets we know something about, life is an oddity in our Solar System. What parts of the planet now, are original – pre-biotic? It's not easy to say. Nor can it be easily asserted that life began on Earth.

It just looks more like life had its origin outside our planetary system or alternatively (and for practical purposes this amounts to the same thing) its origin on a quite different early Earth has been so completely been overwritten like some giant palimpsest that its origin is virtually the same as extraterrestrial.

And then to take another direction, it mightn't have an origin. It might be organised *recursively*, like a Mandelbrot function, and its origins go back *ad infinitum*.

A second argument for the non-origin of life is even more fascinating. Let us remember that we are part of life and we (the human race, the brain, consciousness) are within not outside the life system that we are trying to objectively understand. Our consciousness is part of the living system so try as we might we can't view life objectively. All our thoughts, our senses and our reasoning is a consequence of our evolution. It's like the child trying to describe "home" or "mother" knowing only one home or one mother.

Indeed, this is a thought. Perhaps or suppose there is in the universe no other life bioform which has evolved consciousness. We ourselves have consciousness and can reason but we ourselves are within the system which has come to be and can know no more of it than the captives in Plato's Cave and for the same reason. If this were so then the origin and nature of life would forever remain a genuine mystery. Life is open for the rest of the universe to accommodate it but closed because there is no mind to objectively apprehend it (except, perhaps, God's) and thus there can be no objective, rational account of what life is about. If that is so, and we are the only rational, conscious minds, then the mystery of life would be a genuine and complete mystery with no answer to life's origin and purpose. Indeed no origin and no purpose knowable, because there might be no external mind to "know".

The third anomaly comes about if one is careless about the thermodynamics of living systems. Physics has concluded in the second law of thermodynamics that things tend naturally to go from order to disorder (and increase entropy) unless forced otherwise through work or effort and, provided you do not have over-ambitious expectations of what evolution can achieve, evolutionary theory works, as it must, in accordance with the second law.(2)

Teilhard de Chardin tried to understand the evolution of living systems as undergoing "complexification". If so, this would run counter to the second law which indicates a tendency for the opposite, that things, left to themselves, fall apart and don't build up and don't get more complex.(3)

Prigogine gave the thermodynamic conditions for when there might be a build up, or evolution - "in open systems far from equilibrium". Prigogine's words, the title of his book, "order out of chaos", themselves illustrate the second law; that while work is always needed to bring order *from* chaos, no effort is required to turn order *into* chaos.

There are also examples of "self-organisation" (so Stuart Kauffman) or "emergent qualities" which while not running counter to the second law, are nonetheless increasing order. The cyclic chemical reactions of Belousov-Zhabotinsky variety and the hypercycles of self-organisation shown by Eigen and Shuster are all in some way thought to be clues to how life works, which in the classic work of Jacques Monod is part chance and part necessity (4).

Let us consider the conditions indicated by "in open systems far from equilibrium". All living things, together, on this planet form an open system. Perhaps, if we were to include the Sun it might be thought as a closed system in terms of energy. A closed system would gradually degenerate and *increase* entropy. A system in thermodynamic equilibrium would likewise become more and disordered until moribund.

"Far from thermodynamic equilibrium" means having hot and cold regions, like the Sun and the Earth, in which interesting things can happen as on Earth. Heat will tend to move to the colder regions and by so doing might, as transferred energy, do useful work and *decease* entropy, sustaining life in our case. Whereas no heat will move out of the colder region to the hot. So says the second law of thermodynamics. At thermodynamic equilibrium there are no temperature differences and hence no activity and no work can be got out of the system. Thus plants or animals use the 'energy' and they take it in to live – be it sunlight or proteins. (5) When they stop taking in 'energy' they die – and decay – and reach maximum entropy. No problem there. The problem is what drives the evolution that produced the plant or animal from a previous species or population that was in a certain sense less well evolved? We know what pushes the weed through the tarmac path, but what pushes the species onto the next rung on the evolutionary ladder or up Mount Improbable? (6) – what is the evolutionary equivalent to the chloroplasts that does the work of photosynthesis in plants?

A steam engine with no boiler and cylinder (no Carnot cycle, or chloroplasts) might have a temperature gradient (between the fire-box and the ambient temperature) but if it has no mechanism for utilising the energy as work, it will not turn that work into mechanical motion.

A biological cell is not mechanical. It uses chemicals (such as adenosine triphosphate, ATP) to build macromolecules from monomers, like proteins and nucleotides. To do this it works rather more like a builder building a house than a steam engine utilising energy and converting it into reciprocating motion.

The house builder draws on energy from the local electrical supply; and uses fossil fuels to drive the cement mixers and food to feed the labourers, who literally do the "work". The constructors need sources of energy. They also need raw materials; bricks, tiles, timber, windows and doors. And, this is important, they need the plans or ideas from which to build. These three are needed; usable energy; materials and ideas or information by which to build.

Likewise with a biological system, such as a cell, there is a need for not just energy and the basic building blocks but also for the information which directs what to do with building elements. Cells use energy (and enzymes) to do the work needed to build chemicals and for that you need building blocks *and* information – for you must know what you are going to do before you do it.

References

1.Arthur O Lovejoy *The Great Chain of Being* (1936). The medieval scheme followed through by Lovejoy and envisaging Man well below extraterrestrials such as angels and archangels, but well above spiders and worms, firmly ended in two stops; in 1859 and 1953. What was lost was the marvellous cosmic dimension of the Great Chain idea.

2. Sir Arthur Eddington was in no doubt as to the cruciality of the second law: "If.... your pet theory of the universe is in disagreement with Maxwell's equations - then so much the worse for Maxwell's equations.... But if your theory is found to be against the second law of thermodynamics - I can offer you no hope.." Sir Arthur Eddington, *The Nature of the Physical World* (1929).

3. In fact it was this criticism that Sir Peter Medawar directed at Pierre Teilhard de Chardin's *Phenomenon of Man* (1955) in a savage review article, "Teilhard's belief, enthusiastically shared by Sir Julian Huxley, that evolution flouts or foils the second law of thermodynamics is based on a confusion of thought; and the idea that evolution has a main track or privileged axis is unsupported by scientific evidence."

4 These brief references were taken from *Aspects of Theism* by Colin Price, Mariano Artigas and Philip Clayton (1994) which had limited circulation, but the works mentioned, which can be consulted, are:

I Prigogine and I Stengers *Order out of Chaos* (1984)

J Tyson *The Belousov-Zhabotinsky reaction* (1976)

M Eigen and P Schuster *The hypercycle: a principle of natural self-organisation* (1979)

J Monod *Chance and Necessity* (1975)

S Kaffman *At home in the Universe* (1995)

5. Roger Penrose explains this more fully. I have added the enumeration. "There is a common misconception that the energy supplied by the Sun is what our survival depends upon. This is misleading. [firstly] For that *energy* to be of any use to at all it must be provided in a low-entropy form. [secondly] .. we are fortunate that the Sun is a *hot spot* in any otherwise *cold background*... An energy supply in thermal equilibrium [would be] useless. [thirdly] .. During the day energy reaches the Earth from the Sun, but during the course of the day and night it all goes back again into space..[fourthly] However, what we get from the Sun is in the form of individual photons of high energy.... whereas the energy that goes back into space is in the form of low energy (infra red, which from Planck's equation, $E=hv$ where v is frequency, is less energetic) ..Plants make use of this low-entropy energy in their photosynthesis, thereby reducing their own entropy.. and we eat them or something that eats them." Sir Roger Penrose *The Road to Reality – a complete guide to the Laws of the Universe* (2004) p705

6. "Mount Improbable" i.e. evolution by chance increments, comes from *Climbing Mount Improbable* (1997) by Richard Dawkins which *is* (I think) at odds not so much with the second law of thermodynamics as the laws of probability. Dawkins suggests this improbability can be climbed via a gentler slope, little by little, but with no available work at hand for climbing (only random jolts, on the level) it is the height that makes the ascent improbable, if not impossible, not the slope: the slope is irrelevant. To climb a physical hill against gravity one needs to "do" work (= force x vertical distance). Here, by analogy, for gravitational attraction read, the disordering effect of the second law. But from where comes the applied work from purely random events (with no thermal difference)?

EVOLVING OR UNFOLDING?

I left Part 4 with the supposition that the living system on Earth in order to evolve needs information. In fact, in general, you need low-entropy energy, building materials and patterns on which to build. Plants (which grow from seeds) "unfold". The information or pattern lies within the fertilised seed, which is, in fact, a single cell. We, human beings, also grow from a single fertilised cell, but also we "evolve". Thus, we often talk of "nature and nurture", meaning coming from our genes or as a response to our environment (for instance by switching on particular genes) and so the final result being a bit of both – unfolding and evolving .

Now we can ascribe to the whole, features of the particular. This is the principle of induction. We know what the particulars are, they are plants or animals or like us members of the human race. As to the general, I have called it in previous Parts, the Earth bioform or Life or Nature or evolution, or just the living process.

So I set the question. Is the living process or evolution a thing in itself which evolves and unfolds? And therefore, has it a "family resemblance" (as Wittgenstein might have said) to a species? In which case how does it both evolve and unfold? Or is it more simply just a conglomeration of disparate individuals – variations on a theme of DNA? Though the latter option is a possibility, we examine here the former.

We start at the beginning with the Big Bang. According to the Standard Model, one piece of evidence of which is the cosmic microwave background, now at 2.7 deg Kelvin and which originates some 300,000 years after the Big Bang when the radiation separated into a re-ionised, transparent phase which is still with us today. The Big Bang itself was a singularity at which the equations break down. The earliest moment identifiable is at Planck time, 10^{-43} secs after the Big Bang and 13.7 billion years ago. Then, the temperature was 10^{33} deg K; cooling as the universe was expanding. It is from this cooling of heat energy which sets the direction of the universe. Although the energy in the universe is and has always been constant, neither created not destroyed, during the cooling quarks combined to form protons and neutrons as hydrogen, then by nuclear fusion as helium then in the first three minutes all the light elements to boron. The first stars were formed 13.5 billion years ago. The oldest stars in our Galaxy, the Milky Way, are 10.4 billion years old. Our Sun is a third generation star. And our planet was formed 4.7 billions years ago. Life or the earliest fossils date from 3.8 billion years ago. Life itself depends on the original elements, hydrogen, carbon, oxygen, nitrogen but also some of the heavier metals which came from earlier exploding stars (supernova) (1).

The Universe is expanding (according to Hubble's law) and in that sense evolving, but order is not overall increasing, rather quite the opposite. The early Universe was more ordered (at minimum or near zero entropy) under one united force rather than the four forces of our era, gravity, electro-magnetism and the weak and strong forces within the atom. The unfolding process is governed by the laws of physics in which entropy is increasing. These laws are not like human laws which have to be formulated and enforced but are intrinsic, within the fabric of the Universe. But they have this in common; human laws, even the rules of chess share with the laws of physics that they pre-date the events they purport to govern. Particles *follow* an inverse square law of gravitational attraction. Natural selection, on the other hand, takes from the present a random mutation that has behind it no history and passes it on to the future. That is the difference.

Energy too, is intrinsic, within the Universe. Energy has no origin, being neither created nor destroyed, and was around before Planck time and will be around at the end of the Universe, though not in any form which is usable. Life too, has similarities to this. On Earth, life is almost immortal, though its forms are forever changing as well as its component entities living and dying. And possibly we understand both energy and life with a similar amount of ignorance. (2)

If we truly wanted to understand the life on Earth in terms in which physicists understand the physical universe, we would have to envisage some form of an unfolding of life's varieties and potential rather than an evolving of new forms from random mutations alone.

I confess I don't know how the patterns for evolution or unfolding are held within or without the DNA molecule. (But see (3) for an informed possibility.) I can only suggest there are "seeds within the seeds" (see *recursiveness* in Part 4). And these set the pattern of life on Earth. But I am sure that the standard theory of evolution undervalues the whole complexity of the life process. It has evolution explaining how random changes in the present affect through natural selection the future, but leaves unanswered how present might be dependent on the deep past.

Let me give an example of what I mean from the mathematical world of axiomatic systems, like Euclid's geometry (but more precisely those systems of the Zermelo-Frankel type).

Euclid's axioms are few in number and self-evidently true, such as "only one straight line can be drawn between two points". From these axioms plus some rules such as "things equal to the same thing are equal", interesting theorems, like that of Pythagoras ("the square on the hypotenuse is equal to the sum of the squares on the other two sides") can be deduced. At first it seems as if new information has been invented from basically un-informative axioms, just as evolutionists think something new always evolves in the evolutionary process. After all in both cases the outcome seems to have increased in order, the theorem in acquired knowledge and the species in new adaptations.

But in terms of information nothing new has been made. All that Pythagoras' theorem contains is already contained in the basic axioms. Pythagoras' theorem is simply a novel way of presenting what is already contained in the axioms. This is made plain by, for instance, Carl Hempel (4).

I'd like to suggest that there could here be a similarity with evolution. (5)

Axiomatic systems need not have just a minimum number of simple axioms. The PC computer is a similar deduction system, with a large hierarchy of non-simple axioms including an operating system and programs for this and that; the keyboard input, for instance. The user can get the computer to do nothing which has not been already pre-programmed. So the computer too, shows the same direction of information, even though you might think the computer is doing something new when you use it.

The cells in all living things are similarly pre-programmed, an instance being the triplet codon code. This acts as a program for the gene in selecting the sequence of amino acids to build proteins of which there are some thirty thousand different ones in the human body. Proteins are strings of amino acids, to be fitted and enfolded in a pre-arranged way from a selection of the twenty amino acids that Nature has chosen to use. A typical protein might contain hundreds or thousands of amino acid units.

The codon code is a three symbol code made up of any three of the four DNA bases and in any order. That gives $4^3 =$ 64 distinct arrangements to code for the 20 amino acids, some of which can be duplicated as a safeguard against copying errors. Also included are stops and starts of the gene sequence of amino acids (6).

It is this string of triplet codes that make up each gene (exon, to be precise) and it is the genes and their spacings (introns) that make up the DNA molecule. And, of course, it is the DNA molecules that make up life on Earth.

In conclusion I see every reason to support the supposition (particularly with Penrose's probability argument given in reference 3) that the evolutionary life process on Earth is a profoundly complex, evolving but also unfolding activity; dependent on more information than that which random mutations alone could provide.

References

1. Brian May, Patrick Moore, Chris Lintott, *Bang! - the complete history of the Universe* (2007) p192

2. Peter Atkins doubts whether anyone truly understands what energy is. Entropy, on the other hand, he considers quite straightforward. - see Peter Atkins *Galileo's Finger* (2003) chapters 3 Energy and 4 Entropy.

3. The tricky question of where and how all the information for future evolution is primarily encoded, has a surprisingly simple answer, which I will relate - from Sir Roger Penrose *The Road to*

Reality – a complete guide to the Laws of the Universe (2004) p.764.

We start by considering the Big Bang as having an "extraordinary 'specialness'" as Penrose calls it - in other words the chance of it happening being theoretically near impossible. One calculates the odds as follows (according to Penrose). The odds are 10 to the power of the entropy of the observable universe today, assuming it was a black hole singularity at the time of the Big Bang. Now the entropy of a black hole according to the Beckenstein – Hawking equation is, $S=A/4$ where S is the black hole entropy and A is the area of the black hole in suitable fundamental units. So $S=$ 10 raised to the power of 123 or 10^{123} (why 123? - see text) – a very large number and the likelihood of the Big Bang becomes 1 against 10 raised to this power – an even larger impossible number, viz 10 to the 10^{123}

Penrose says concerning our planet and life itself "One can estimate the entire Solar System, including its living inhabitants could have been created from the random collision of particles and radiation with a probability of 10 raised to the power of, say, 60 i.e. 10^{60}. This is utter "chicken feed" by comparison to [the previous figure of 10 to the 10^{123}] needed for the Big Bang."

What Penrose is saying (if I read him aright): The Creator (Penrose's choice of word) could have given himself a much easier job by producing Life directly without going to the bother of a Big Bang. In which case the Creator (or a singular event, you don't have to be religious, here) could have packed the life system with enough information to power the living evolutionary process until the end of Time and it *would have still been a much more likely event than the Big Bang, the origin of the Universe.* I just need to add, of course, as we all know, the Universe and Life, however seemingly impossible, happened!

4. Carl Hempel's numerous works on deduction helped to clarify this point for me.

5...following on: they (the evolutionists) also often point out, as evidence of evolution, that the direction of the fossil record indicates a progression from single-celled, to multi-celled, to tetrapods and so on to the mammals and primates. But Nature could also be like the geometrician working out, first, the simple theorems, QED, before advancing to the more complex; though, in fact, they all have an equal status as true deductions from the basic axioms, irrespective whether they are simple or complex. Neither can any one theorem be newer or older than any other.

6. Peter Atkins' account includes a twenty-first amino acid, selenocysteine, which uses the relatively rare element, selenium. The triplet code is TGA which is already used as STOP. One can see the point. The gene seems to be saying if available use selenocysteine; if not stop building the protein at this point. *Op cit* page 67 I could add that the four paired bases mentioned in the main text are Adenine and Thymine; Guanine and Cytosine, more usually known by just their capital letters.

The nature of Nature – Part 6

UNFOLDING OR SUCKING?

Every age has added knowledge of the content of the natural world, the world of reproducing and growing things as opposed to the world of inanimate objects, for instance by naming and recording species.

The problem has been not the content but the interpretation of the content. The first obvious interpretation was that it was the work of the Creator. Later, the secular idea that the living world arose by chance and that it also to this day proceeds by a natural (as opposed to an interventionist) process of natural selection, on essential random changes or mutations, as described by Charles Darwin, 150 years ago in the *Origin of Species*.

We are going with the notion that the living world is a creation, but I am leaving as other subjects, such as religious belief and theology, the nature of the means which caused or brought about this creation, the Creator.

The world has two features that are fairly indisputable - it is not static and time moves in one direction only. So is the world the product an evolving creation or an unfolding creation?

In 1943 Erwin Schrödinger, already renowned as a quantum physicist for his formulation of the wave equation in 1926, gave a lecture at the Institute of Advanced Study at Trinity College, Dublin which was later published as a small book, *What is Life?* (1944). As an eminent scientist but no biologist, his analysis and insights proved to be original and seminal. Ten years on James Watson was to say how influential Schrödinger's thought was when the full significance of the structure of what was thought to be an uninteresting and inert substance, namely DNA, was published in 1953.

Schrödinger's thoughts on the physics of life were also interesting and original. He proved to be both right and wrong in his speculations on the thermodynamics of living systems for although he was convinced unlike the majority of his contempories that the laws of physics must apply to biology and biology wasn't therefore a special case, his analysis of how living things maintained (and increased order) "by sucking orderliness from the environment" is mainly wrong and unnecessarily over-complicated.

Organisms, Schrödinger said, feed on "negative entropy" by "drinking in orderliness from a suitable environment". Organisms are fuelled, certainly, by food, which is an energy input (and which also decreases the entropy of the organism) and by water and oxygen. It is these that maintain the organism in order. And that order will maintain, like Canute against the tide, so long as the food supply, be it protein or sunlight, is kept up or until death intervenes. At which time the order begins to be lost and the organism begins to disintegrate or before that point is finally reached that low-entropy order might be passed on usefully as food to maintain a further individual's life and reduce its entropy.

Even plant remains, because they are still of low-entropy, can pass on energy transfers millions of years later as coal and oil before they finally reduce to maximum entropy, where no further energy extraction is possible. So living things do have the same energy requirements as any other physical energy converter, including steam engines and electric motors.

We have come full circle since René Descartes postulated that animals were mere machines and we, human beings, machines with souls through to those who maintained a distinction like Henri Bergson who described an *élan vital* in his 1907 work *Creation Evolution*. We can now envisage living systems as machine-like or rather our human machines, steam engines and electric motors, as life-like. This, in the sense that an energy converter, be it from a photosynthetic plant cell converting solar radiation into chemical energy or a steam engine converting low-entropy coal into heat into steam into motion, has similar components. These are the basic elements of a machine; fuel input and another form of energy as output with a temperature difference between the two. An actual difference is that human heat engines have an immediate temperature difference like that between the firebox and the ambient temperature in the case of a steam locomotive whereas, in photosynthesis or the equivalent in the animal mitochondria through the judicious application of enzymes, the necessary chemical reactions occur at room temperature.

But behind all life lies the fact that living systems are not closed systems but rely on low-entropy solar radiation from a hot source, namely the Sun transferred through a colder background, the Earth. From this basic high energy, low-entropy source comes the start of the energy-transference chain which we rely upon – the food chain. Plants convert about 4% of the total solar radiation falling on the Earth, which amounts to some 2×10^{23} joules /year. Although we ultimately depend on the higher orderliness of the Sun we also "suck" orderliness from any fuel or food as secondary sources that has a high calorific value and a low entropy. We don't eat ashes because they are at near maximum entropy and have a negligible calorific value!

The point is that there needs be something else other than just an energy transfer from a hot, low entropy source to the receiving end. With nothing else the Sun's radiation would just end up as heat. But it doesn't. 4% of it turns up as available energy, thanks to photosynthesis. There needs be mechanisms, or structures or chemical cycles to affect the transfer from one form of energy into another. It can become fuel or food only if there is a process processing it.

There are some features in the very origin of living systems, such as how the first self-replicating molecules arose or how the first chlorophyllic cells, which is the basis of the energy chain of life, came about. These are inexplicable jumps, going the wrong way in the strict thermodynamic order of things.

Roger Penrose describes these "inexplicable jumps" (as I call them) in terms of the likelihood or not of finding very small volumes ("boxes") in a general phase space as events change with time. In brackets [ ] I have exemplified the case of the Earth before and after the evidential advent of life.

"...In other words, the entropy of the system will indeed get larger and larger as time marches forward. Once x [say, an event in Earth's history] finds its way into a box [i.e. a discrete volume of phase space] with a certain entropy measure, [such as that obtained before the emergence of life on Earth, say, 4 billion years ago] it becomes overwhelmingly unlikely that, in any sensible period of time, it can find itself back in a box of significantly smaller entropy [such as a lower entropy change equivalent to the organisation needed for the initial formation of the primary elements of life, such as the emergence of self-replicating molecules and chlorophyllic cells to begin the Solar energy chain, for mitochondria, prokaryotic cells, etc to make use of] would mean finding an absurdly tinier volume [of phase space] and the odds are immensely against it. Think of the example we have been considering, [see original text for the calculations] and the absolutely stupendous reduction in phase-space volume [broadly, the chances against it happening] that would accompany a very modest reduction in entropy, [and the coming about of a primary element of life, say, the chlorophyllic cycle/cell would not require a "modest reduction in entropy"! - it would involve a large increase in order/reduction in entropy] *owing to the logarithm in Boltzmann's formula and the small size of Boltzmann's constant."* [which together make plain Penrose's first point, the overwhelming probability is that *the entropy of the system will indeed get larger and larger as time marches forward.*]

Roger Penrose *The Road to Reality – a complete guide to the Laws of the Universe* (2004) p697

The above analysis seems to rule out any possibility of life on Earth being the result of accident or chance mutation. Indeed, it seems to rule out any possibility of life on Earth happening at all! But this we know is wrong! Let us be clear. What we have ruled out is not the existence of suitable energy sources to drive the life system - there's plenty of available energy, for instance photons from the Sun, but without a low entropy means on Earth to get any life system kick-started, the only source of low entropy radiation, the Sun, would all go into just illuminating the Earth and heating it up. Nothing else. So from where did Life come?

Might life have had an extra-terrestrial origin, for instance? One could bring to mind the "panspermia" theory, revised and associated with the names of Fred Hoyle and Chandra Wickramasinghe. And one could thermodynamically open up the Solar system from being an essentially a closed system to being open but the immensity of the spatial scale would place a successful arrival on Earth of a successful seed system no more likely than the spontaneous generation of self-replicating molecules or chlorophyllic cells which have already been ruled out.

But secondly, Penrose's analysis would also indicate that the "bit by bit" evolution of the mammalian eye or the bacterial flagellum etc and similar evolutionary events, far from being more likely to happen, because they are "bit by bit" (as Richard Dawkins famously tried to argue with his analogy of climbing the easy way up Mount Improbable) are *less* likely to happen as this too involves a series of smaller and smaller "Penrose boxes".

The remaining possibility is that the information is already encoded and is already accounted for in the entropy figure of the pre-biotic Earth, which we took as that measured 4 billion years ago.

In the same way the physical forces of geology and meteorology, in the same sense, unfold the Earth's geophysical appearance, driven, among other factors, by internal sources of nuclear activity, to which the biosphere eventually makes substantial contributions.

The nature of Nature - part 7

THE PENROSE EGG PROBLEM

The burden of these papers has rested on the supposition that the structure of living forms - the biosphere - is more than can be accounted for by an accidental sequence of purely random events. The biosphere is unique as far as we know. There are stars and planets aplenty made up of the elements of the periodic table but none as far as we know with the products of life as its elements, the macromolecules such as the polysaccharides, polypeptides and proteins. Earth has a surface covered deeply with the products of living things from sedimentary stone to carboniferous coal. How did it all come about? What is the nature of Nature?

I want to use Roger Penrose's work to make two points; to emphasize his calculations on the extraordinary unlikelihood of there being life at all from the chance concatenations of atoms and molecules, and what it means in the light that there is life (!) But first to answer the question if life "unfolded" (rather than evolved from nothing) from where did the information come?

from page 36

3. *The tricky question of where and how all the information for future evolution is primarily encoded, has a surprisingly simple answer, which I will relate - from Sir Roger Penrose The Road to Reality – a complete guide to the Laws of the Universe (2004) p.764.*

We start by considering the Big Bang as having an "extraordinary 'specialness'" as Penrose calls it - in other words the chance of it happening being theoretically near impossible. One calculates the odds as follows (according to Penrose). The odds are 10 to the power of the entropy of the observable universe today, assuming it was a black hole singularity at the time of the Big Bang. Now the entropy of a black *hole according to the Beckenstein – Hawking equation is, $S = A/4$ where S is the black hole entropy and A is the area of the black hole in suitably natural or fundamental units. So $S = 10$ raised to the power of 123 or 10^{123} (why 123? - see the original text) - a very large number and the likelihood of the Big Bang becomes 1 against 10 raised to this power – an even larger impossible number, viz 10 to the 10^{123}*

Penrose says concerning our planet and life itself: *"One can estimate the entire Solar System, including its living inhabitants could have been created from the random collision of particles and radiation with a probability of 10 raised to the power of, say, 60 i.e. 10^{60}. This is utter "chicken feed" by comparison to [the previous figure of 10 to the 10^{123}] needed for the Big Bang."*

What Penrose is saying (if I read him aright). The Creator (Penrose's choice of word) could have given himself a much easier job by producing Life directly without going to the bother of a Big Bang. In which case the Creator (or a singular event, you don't have to be religious, here) could have packed the life system with enough information to power the living evolutionary process until the end of Time and it would have still been a much more likely event than the Big Bang, the origin of the Universe. I just need to add, of course, as we all know, the Universe and Life, however seemingly impossible, happened!

Suppose we take one of Penrose's enormously large figures – call it the **S Factor** – this now represents the odds against there being Life anywhere in the Universe. But there is life. Somehow the odds have been nudged by at least a factor of the S Factor. The S Factor now becomes a measure of a shift in the original state of affairs which we either had no knowledge of nor had we anticipated it.

The S Factor becomes a measure of how much lower the original entropy of the Universe must have been in order that we now have life.

We don't have a clear understanding of the nature of life. Most people suppose it is an emergent property of certain chemicals at a certain level of complexity, but equally it could be that certain chemicals at a certain level of complexity, (polypeptides and nucleotides, etc) are manifestations of there being life, which is quite different!

Let's pursue this by looking at another piece of Penrose's writing.

From *Cycles of Time* p26 by Roger Penrose *The Egg on the Table*:

"Let us consider the egg perched on the edge of a table [mentioned earlier]about to fall off and smash on the floor below [and therefore soon to be at near-max entropy]

The entropy-raising process of the egg rolling off the table and smashing is enormously favoured probabilistically [eggs are more likely to smash than not] provided we are prepared to assume that the egg in a very low entropy state [ie highly ordered] of being perched unbroken at the edge of the table. the puzzle of the second law is not the raising of the entropy as it smashes ; the puzzle lies in the event itself ie the question of how the egg found itself in this extremely low entropy state in the first place. The second law tells us that it must have arrived at this very improbable state through a sequence of other states even more improbable prior to this and getting more so the further back we go.

[At this point I would like to refer the reader to the piece on 'Penrose boxes' p42 and p697 in the original text. As life evolves, thermodynamically boxes of phase space get tinier and tinier and the odds at finding anyone are immensely against it. To bring the odds to at least even requires the original entropy to be lowered by, **S Factor**, at least.]

Penrose continues

There are basically two things to explain. One is the question of how the egg got on the table, the other is how the low entropy state of the egg itself came about. Indeed, the material of the [hen's] egg has been superbly organized into a perfect package for the [future] chick.

But let us start with what may seem to be the easier part of the problem namely how the egg [got on the table]. The likely answer is that some person put it there, perhaps a little carelessly, but human intervention is the probable cause.

There is clearly a lot of highly organised structure in a functioning human being which suggests low entropy and the placing of the egg on the table would have taken only a very little from a huge reservoir of low entropy...consisting of a well fed human being surrounded by an oxygen-laden atmosphere.

The situation with the egg itself is somewhat similar, in that the egg's highly organised structure... is very much part of the grand scheme of things that keeps life going on this planet.

The entire fabric of life on Earth requires the maintaining of a profound and subtle organization, which involves entropy being kept at a low level."

How is it that the "fabric of life" is kept not only at a low level of entropy but grows? This puzzled Erwin Schrödinger in his book, *What is Life?* (1944)

Schrödinger's paradox is why do living systems "grow", which they do by apparently imbibing order out of the environment, when closed systems ought to get more disordered?

This is the opposite to entropy which Schrödinger calls *negative entropy or neg-entropy* i.e. structural information through somehow order is built up.

And we see how, with the egg on the table. For in life on Earth we have "a large reservoir" held at low entropy (largely due to the effects of photosynthesis)., which doesn't so much "suck" or "unfold" but *"gives" to another generation a loan of energy to kick-start new life.* There is "food" in the egg and perhaps the promise of heat from the hen. The paradox is simply solved by an ever-so-slight decrease in the vast reservoir of usable energy that is the Earth bio-system to maintain the egg at an ordered state of low entropy (remembering that low entropy means greater order). Thus the one balances the other. Living systems are not closed systems. The gain in order of the egg is at the expense of some slight disorder in the vast bio-system, (amid which is a hen pecking at chicken feed and burning it up as food).

So where did this vast reservoir of usable energy come from? It is made up of the vegetation covering the Earth, and the coal and oil beneath. All these things are burnable and burnability, implies low entropy i.e. ordered structure, the detritus of once living things. Only the Earth with life has on it and in it burnable things. All of which ultimately derives from the Sun, as products of photosynthesis or as products of things which themselves depend on photosynthesis.

Roger Penrose puts the matter concisely in both the books already mentioned.

During the day photons from the hot Sun arrive on Earth and during the day and night are reflected back into cold, dark space with a lower energy, that is longer wavelength.

The Earth does not grow any hotter over time therefore there is an energy balance, but the energy coming in from the Sun carries considerably lower entropy than that returning to space. By using photosynthesis, green plants have found a way of converting this energy difference into low entropy substances, such as sugars, using CO_2 [and by splitting water] returning it as O_2. When animals eat vegetables (or animals which eat vegetables) they use this source of low-level entropy to keep their own entropy low.

But how did it all start?

We can, in imagination, follow backwards in time the "vast reservoir" of the living Earth and see it getting smaller and smaller as we reach back to the very advent of life on Earth, say somewhere between 3.5 and 4 billion years ago; to the very first chlorophyllic cell; the first evidence of life on Earth; the very beginning of the energy food chain.

There is our egg, one of a long line of eggs, alas no longer on a table, the very first cell (eggs are single cells) from which all living things came from and there's nothing else; no "vast reservoir of low entropy," (as Penrose calls it; or the neg-entropy of Schrödinger) suddenly Schrödinger's paradox comes alive again. At this point we might well ask, how did it all begin – what came first? There's nothing to "kick-start" the first element of life be it a strand of RNA or a chloroplast; no disequilibrium of energy; all a stray photon can do is heat up a bit of the ground. There's no photosynthetic process and if there was where did it come from? The Sun, unlike the "reservoir" on the later Earth, is no source of structural information (such as that needed to make an egg) even though both are held at low entropy. The Sun supplies the energy, the information must come from elsewhere.

The origin of all living things

The general view is that all *present* living things can be traced back to three domains; Archaea, Bacteria and Eucaryota. (All animals, plants and fungi comprise a small part of the last named.) Some have seen all things traced back to LUCA (the Last Universal Common Ancestor). Two things seem common to all living things; a replicating macromolecule like DNA (or possibly RNA) and an original cell, from which all cells derive by cell division.

(The material constituents of cells are not all contained in the DNA.) Thus we seek two origins, that of nucleotides and that of cells.

But origins are tricky! Penrose says that the egg *must have arrived at this very improbable state through a sequence of other states even more improbable prior to this and getting more so the further back we go.*

Remember the golden hamster story. All present-day pet hamsters derive from a litter discovered in the Syrian desert by Israel Aharoni in the 1930s. No one suggests before that there were no golden hamsters. Here is another story. When the composer-lyricist, Sammy Cahn was asked, "Which comes first, the words or the music?" His answer was, "It's usually a telephone call!" It's neither this nor that but something quite different. There is a fallacy which runs, "because, we think we know how things are now we assume this is how they began."

Two accounts of how aeroplanes are made

Things don't come about from nothing (*ex nihilo nihil fit*). The Wright brothers "built bicycles but dreamed of making flying machines" so one of them once said. They started with bicycles and ideas (or possibly ideas and bicycles). The "egg on the table" was their first plane, the Flyer at Kitty Hawk, North Carolina in 1903. The vast reservoir, which both gave thermodynamic support and structural information, drew on an information pool of metal & woodwork engineering and engineering know-how, and lots of other things, not the least of which was, the study of bird flight. Nor should we forget human ingenuity, which was driven, perhaps least of all by many wholesome meals, including (I hope) vegetables. The Kitty Hawk Flyer, from being "the egg on the table" now becomes part of vast reservoir. The knowledge and information it holds becomes held in the general pool, while its actual parts are freely available for recycling – but actually end up in the Smithsonian Institution. The *idea* of Kitty Hawk Flyer and others begat further aeroplanes and although there is no genetic relationship (of course) there is a virtual one.

The second account is much less creditable and asks us to believe that life on this planet came about by chance and was refuted as such by Fred Hoyle.

Fred Hoyle said the chance of the random emergence of even the simplest cell was about as likely as "a tornado sweeping through a junk yard and assembling a Boeing 747".

So, if you if you believe the Miller-Urey experiments (p17) emulate the way life started, you might also think this is the second way of making an aeroplane.

But I'm with Hoyle, who, even though an atheist, saw the Universe as too intricate to be random when he discussed this in The Intelligent Universe (1983.)

We are back to our starting point – to what Penrose describes as the "extraordinary specialness" of the Big Bang and this was my comment (p37). "The Creator (Penrose's choice of word) with these odds could have packed the life system with enough information to power it to the end of Time and *it **would** have still been a much more likely event than the Big Bang.*"

A variation on the Sherlock Holmes adage

If the improbable verges on the impossible and still happens, then there is something amiss in your account of the grand scheme of things.

This is where S Factor comes in – the bias held somewhere which allows the impossible odds to happen.

Put it another way. If you looked at life dispassionately and "back-engineered" the whole project you'd need more credit in the bank (i.e. a factor of the order of S Factor) than the "something from nothing" evolutionists suppose. The reason why the Miller-Urey experiment is no answer to how life originated is because (among other things) there is in the so-called "primeval soup" no thermodynamic disequilibrium.

Remember, under the second law of thermodynamics, if there is no money in the bank, order can't increase   (de Chardin supposed it could (see p27 ref 2 &3)) nor are you allowed to fall into debt, that is, go below absolute zero.

The energy in the universe becomes increasingly unavailable for useful work the closer the system approaches absolute zero and therefore as time goes on, it becomes increasingly more difficult to introduce complexity and variety.

There needs to be (somewhere in the Universe) a vast reservoir of low entropy information to make all life on Earth the equivalent of Penrose's egg on a table.

References

The quotations are taken from Roger Penrose *The Laws of the Universe* (2004) ; *The Cycles of Time* (2010) and my own *The nature of Nature* (2011) in this present volume.

THE NATURE OF NATURE

GÖDEL
AND OTHER
TWISTS OF
LIFE

THE NATURE OF NATURE

Making use of

Gödel's theorem

Gödel's incompleteness theorem (1) is often mentioned in the literature on science and religion, sometimes (and vaguely) to indicate an inability of the human mind to grasp anything completely or less generally to indicate the limitations of science especially with regard to religious questions.

All this does no justice to science or religion and rather makes light of the work of Gödel which proved to be one of the most astounding conclusions in 20th century mathematics.

The Cretan who said "all Cretans are liars"

Gödel's paper was a result of the failure of the Russell-Whitehead project, *Principia Mathematica* (CUP, 1910-13) to cope with the presence of certain paradoxes in order to logically derive the whole of mathematics from an axiomatic basis.

Soon after the project began Russell met with a type of irritating paradox to do with "the class of classes which are not members of themselves" - which, it turns out, both are members and are not members. Most classes are not members themselves. The class of vegetables is not a vegetable, but the class of, for instance, ideas, is also an idea.

The difficulty is similar, in fact, to the puzzle St Paul told Titus about "a true witness" that "*the Cretan who said 'all Cretans are liars'* ". The trouble here is "all" - for it includes the Cretan himself; so that the statement is untrue if they are all liars, but equally. if he is a true witness that they are all liars, the true Cretan witness is also a liar.

Russell at first thought that he could simply (and arbitrarily) eliminate such statements by saying that a statement should not be able to include or refer to itself.

"Never glad confident morning again!"

But when he told Whitehead about this, Whitehead knew better. Quoting from Browning's poem *The Lost Leader*, he remarked, sagely, "Never glad confident morning again!" And such proved to be the case. The problem of defining "the class of classes which are not members of themselves" could be avoided but not eliminated. And so things remained until 1931 when Gödel published his paper.

It might have been that Gödel found a new way round the problem but instead his proof emphasised the problem and gave a real and central prominence to it. Gödel proved that at least one paradox would inevitably arise and no way it could be avoided either by prohibition or by ingenuity. Why? - Because nothing in this Universe is ever complete in itself.

Metamathematics

What Gödel did was construct a parallel or metamathematical system, carefully mapping and labelling each element in an analysable way. In fact he used prime numbers as labels so that each composite (or theorem) was a number whose factors were uniquely prime numbers and referred directly to the axioms, also labelled as prime numbers.

By such labelling Gödel showed that a paradox would arise

and could not be eliminated.

Such systems as Russell and Whitehead's which began with self-evident axioms ("things equal to the same thing are equal" and so on) would always contain theorems which were true but could not be proved.

More than such a system would always contain at least one theorem, which is true but leads to inconsistency. Therefore the described system will have to remain complete but inconsistent or incomplete, but consistent. Never complete and consistent. Adding another axiom to get round the contradiction or incompleteness would simply lead eventually to another paradox.

Gödel's theorem remains one of the great surprises of the 20th century mathematics.

Let me say right away that the incompleteness theorem is strictly a limitation only on mathematical axiomatic systems of the sort and complexity envisaged by Russell and Whitehead. I have also presented the idea that there might be a generalised form.

Quasi-axiomatic systems

But quasi-axiomatic systems also lie at the basis of many enterprises including some in ethics and theology. Spinoza (as well as Russell) was impressed by the power and elegance of the axiomatic system, typically Euclid's geometry - where, by starting with some extremely obvious axioms such as "a line has extension but no breadth" etc one can derive the far from obvious theorem of Pythagoras - that the square on the hypotenuse is equal to the sum of the squares on the other two sides. Spinoza's Ethics is set out axiomatically. Theologies are often axiomatically based - God is One - He is Perfection - and catechisms - The chief end of Man is to glorify God.

Much of the ordering of human society, as in natural law and constitutional law is tantamount to being axiomatic. Both try to contain the whole of human behaviour within a given framework. The American Constitution is an example

Artificial Intelligence, AI

As well, there is its use in Artificial Intelligence. The programming of a computer is like an axiomatic system. The computer can't do more than what is in the program.

And (and perhaps here is the rub) the program can not contain more than there is in the human mind. Indeed the program can not contain the human mind, any more than Russell and Whitehead's *Principia* can contain all of mathematics.

Beware of theories which hope to explain everything. While some of these limitations and defects are no doubt true this is not to say they are true because of Gödel's theorem.

Some things however can be said. Many ideologies, philosophies and theories purport, for various reasons to explain or account for everything. Gödel's theorem in its generalised form demonstrates that even closed systems would have to contain "loose ends". Consider, for instance, the need to add Amendments to the American Constitution; sometimes indeed as contradictions, as in George Orwell's *Animal Farm*: All animals are equal - *but some animals are more equal than others.* Which is also to note that systems of thought which are Gödelianly complete will inevitably contain at least one contradiction.

Beware, therefore, of Greeks bearing one theory to explain everything.

Möbius strip

A paradox can be envisaged as a Möbius strip. The inside being "true" and the outside "false" with the twist being both true and false. (See Note) As a whole the twist makes sense, but looking only at the inside or the outside the twist is paradoxical, belonging not entirely to either. Gödel reminds us that the Universe is essentially One. It is a human expectation that we can divide the whole into descriptions, names or explanations, entire in themselves.

What we are in effect doing by naming or explaining etc is making a "closure" (to use Hilary Lawson's term (2)) and either by dint of our own finitude or for conversational convenience truncating into a particular that which is an integrated whole - the Universe.

Paradoxes ought to remind us that we are only ever looking at a portion of the picture

"Loose ends" and belly buttons

Why should Gödel's theorem arise? No one has ever analysed anything in the natural world quite so exactly as Gödel. So we don't know whether there is a general case to be had and everything not just mathematics is ultimately connected and can't be subdivided. Nothing, it seems, can be absolutely complete, there is always somewhere a "loose end" - an *omphalos*, a navel - which in fact makes it complete - that's the paradox. Certainly we have many "loose ends" in both science and theology. The question is: are they really Gödelian "loose ends" or simply present-day unknowns?

"Loose ends" and scientific unknowns

Here are some.

a. The Big Bang theory begins with a loose end. Energy. Unlike space, time and matter, it is neither created nor destroyed in our Universe, yet it is present.

b. Samuel Butler observed in 1887, "Life is two and two making five". There always seems to be more to life than we can account for. Does this indicate that our understanding of living systems (because we are ourselves are living) will forever be incomplete in a Gödelian sense?

c. Stuart Kauffman (3) suggests for the origin of macro-molecules like DNA and RNA "a self-replicating system capable of at least one thermodynamic work cycle" is needed as an "autonomous agent" to start things off. Another "loose end"? - not the "autonomous agent" but the need for an external agency in the explanation.

d. Will the question of the nature and origin of human consciousness also prove to be "undecideable" or circular - because, like the Cretan's assertion, it is us who are conscious who are asking the question? Perhaps even, because we are conscious, that we want to ask the question. Now that is circular!

e. All living things have natural *omphaloi* – manufactured items usually have a maker's mark which like an *omphalos* paradoxically are part of yet not functionally needed for the manufactured item.

Gödel's contribution -- ultimate questions might well be "undecideable"

Russell's thought, in the early years of the 20th century, ran through, mathematics, logic, language to philosophy. His philosophical objective was the hope that what we truly know, knowledge, would emerge naturally from coherent language, if the foundations of coherent language could be proved to be logical, having already proved that mathematics (itself indubitable) was a species of logic. The hope, therefore was, that discourse on or about the empirical world (and its problems) could be opened up to logical analysis and sorted out. Gödel at least brought things down to earth - life could not be quite that simple. If it was there would be no need for religion!

References

1. Gödel's incompleteness theorem is found as Proposition V1 in *On Formally Undecideable Propositions in* Principia Mathematica *and Related Systems, I Monatshefte fur Mathematik und Physik* vol 38 (1931) pp173-198

This origin paper is difficult and is best read in conjunction with the following, particularly the first, to get even a flavour of what Gödel is about.

E Nagel & JR Newman *Gödel's Proof* (1958)

Roger Penrose *Shadows of the Mind* (1995)

Rudy Rucker *Infinity and the Mind* (1982)

2. Hilary Lawson *Closures - a story of everything* (2001).

3. Stuart Kauffman *Investigations* (2000)

Note- More precisely, if every false statement is written on one side of a paper strip whose length is greater than the width and every true statement on the other side, then a universe of discourse could be made by joining both ends with a twist in the third dimension. The twist (the only element in the third dimension) represents "All statements are either true or false" which is true except for the statement itself whose truth or falsity is undefinable and has to be sought from a higher universe - for instance on the authority of Aristotle to give an example.

First published in the *Bulletin of the Society of Ordained Scientists*

Colin Price © 2004

Note on the generalised theorem.

The "*omphalos*" as in the theory of P H Gosse, in a book of that name, published in 1857, was discussed in an MPhil thesis (Hull 1979). The presence of an *omphalos* indicates that things are seldom complete in themselves, but admit to a "loose end" - a paradoxical element, relating the object to the Universe in general but strictly speaking, like an umbilical chord, being in both camps at the same time. In this case mother and offspring. All plants and animals are related to the ancestral origin of living things - the first dividable cells and the first self-replicating macro-molecule. An equivalent is "loose end" which is my term for man-made objects which relate an object to its human origin - typically to a maker's mark which strictly speaking is not a functioning part of the item. Like Gödel's paradoxical element the maker's mark is an *omphalos* which displays the incompleteness of the item and reminds us that its origin lies elsewhere, in the factory.

But if you put the object and the factory together there is still a loose end; in this case the human race.

The generalised form of Gödel's theorem was suggested and explained in a PhD thesis as Appendix A (Bradford 1984). According to the general theory, arithmetic is not unique in being incomplete - other things too might be incomplete – only in the case of axiomatic systems has the incompleteness been rigorously established. Gödel was able to precisely determine the incompleteness because the peculiar nature of such systems made such an analysis possible. The general conception is that other systems too are similar culminating in the Universe being paradoxically incomplete and yet containing everything. The question of its origin lies beyond this or that answer; it is technically "undecideable". So too are such self-referential questions such as "what is the Universe?".

Gödel's theorem was certainly unexpected, but for the most part, can, even in mathematics be safely put to one side. But its real significance is that things can never be finally untangled. Most things are part of something else. Even mathematics, the great stand-alone edifice is never ending and therefore never complete while there is a human mind around that can supply fresh truths, unforeseen by the axioms.

Gödel's theorem[1], the Omphalos conjecture[2] and the Beatific Vision

Clouds of unknowing

All the main religions of the world have a special place for their own mystics; who hold, more often than not, more in common with each other than with their fellow religionists. In contrast to prayer which is communication with the divine, mysticism is in some sense being "absorbed" in the divine, which, though Athanasius put it rather grandly as, "For He was made man that we might be made God" is actually rather scary.

Often mystics try to describe the experience as a "Oneness" but more often than not they fall dumb at trying to express the inexpressible, which is, stubbornly ineffable. Here, Wm Blake manages better.

> To see a World in a Grain of Sand
> And a Heaven in a Wild Flower,
> Hold Infinity in the palm of your hand
> And Eternity in an hour.
>
> *Auguries of Innocence*

I take this as an experience of Oneness. The Beatific Vision also actually implies a simple form of Oneness; to reverse the words, Vision is a form of knowledge (gnosis); joining with blessedness (makaroi) or happiness (eudynomia). This is a oneness between *thought* and *feelings*; nominally quite separate. For instance, to give an example, Archimedes, in his bath, thought of "displacement", then ran down the street in high emotion, crying, "Eureka!". If he had been thinking at the same time he would have paused and realised he was naked. I rest my case!

William Blake describes, very adequately a Oneness of scale; a Oneness of size between the world and a grain of sand. A Oneness of beauty, to see there would be no difference between the beauty of Heaven and that of a wild flower.

But this Oneness is not obvious in a world that is naturally particulated, itemised, even fractured. Maybe this is the price we pay for language, thought and reason. We name, itemise and fragment the world in order to live in it. "Tree" is the tree we see. We don't see its connection in the soil; its connection with sunlight through the leaves.

To take an example. A mushroom, looks somewhat "treelike" and that's how we think of it. But the mushroom is just the fruiting body; the mycellia might cover acres underground. It's just not that easy to say where a tree begins and ends, even the fungi might be in a symbiotic relationship. At the molecular level there are connections with sky and earth. As an energy system the connections go as far as the Sun. As a biological system the connections go back in time through genomic relationships to the first cell on Earth and before that to the first self-replicating macromolecule.

These are connections but not the connections of Oneness experienced by mystics.

Eternity not just "goes on forever, especially towards the end ", as Woody Allen puts it; eternity is not "everness". Eternity & Time are not species of each other. Eternity is not just past, present and future all joined up.

Eternity is time, "but not as we know it, Jim", (that's a Star Trek quote).

Mystics have their own understanding of terms such as Time, Eternity, Love, Peace, Tranquillity, Holiness, Glory. **God** is Everything, but everything is not **God**.

How does the mystic see "oneness"? We just don't know. Our language fractures the world into nouns and verbs, tenses and empty words. To speak requires thought; to think requires words or if not words, ideas or if not ideas, feelings. So what is it that separates me from the external world? Does the mystic lose sight of "selfness" as well, when all becomes absorbed in the "One"?

Let's start at a far simpler level. Jonathan Sacks, whom I admire greatly, says in *The Great Partnership* (2011) this, about science and religion, "science takes things apart to see how they work. Religion puts things together to see what they mean."

The trouble is science and religion have never been equal partners.

Frazer in *The Golden Bough* saw the tapestry of human history moving through myth, religion and science. True, each point in history contained varying amounts of all three, even though the upward trend was from feelings towards reason; the right side of the brain to left; the yin to the yang and so on.

Let's get even simpler. Pythragoras' music scales had mathematical intervals based on the ratio of whole numbers. Octaves a ratio of 2:1; then there were fifths based on the ratio of 3:2. So why did J S Bach need to write music for "The Well-Tempered Clavier? Why? Because the science of sound needed the temperament of humanity to go from sound to music. Without either it would be sound without music; music without sound.

And this surely is the point we are striving towards. Science can't see how things work without taking them apart and religion can't see what they mean, without putting things together. But without things being capable of being put together they wouldn't be capable of being taken apart.

If you think this sounds circular and slightly paradoxical then it's time I introduced you to Gödel's theorem! But before that let me try and summarise. On the one hand we have an experience of the Beatific Vision where the mystic is led into the mystery of all-existence; absorbed into the Oneness of creation; surrounded by the Beauty of Holiness; the Glory (shakina) of the Lord. "On the other hand" (to quote Topal from *Fiddler on the Roof*) we have our view of the sensible world; a world of beauty, certainly; a world of variety and fruitfulness; a world of things, trees, leaves and mushrooms; an infinity of possibilities from galaxies to atoms and atoms to quarks - and it all makes sense. So where does that leave the mystic's point of view? Is it but a mental aberration in a material world; a breakdown of neurological processes in an otherwise calm sea of rationality? Would we be throwing out the "baby with the bath water" if we refused Jonathan Sack's suggestion of a Great Partnership?

But before I answer that I have to mention something about Axiomatic Systems like Euclid's geometry or indeed the American Constitution.

Axiomatic Systems

The classic example is Euclid's geometry. A system of true statements (theorems) that are logically derived ("deduced") from self-event basic statements called "axioms" Generally few in number and independent, axioms are, such as, "if A=B and B=C then A=C" and simple definitions "a line can be drawn from any point to any point". The remarkable thing about Euclid's system is that from a few self-evident axioms a quite non-obvious theorem like that of Pythagoras (Oh him again!) can be proved true, so that you can truthfully say if the axioms are true, then it is true "that the square on the hypotenuse is equal to the sum of the squares on the other two sides". What is remarkable is not that true things add up to true things (that's what you expect Truth to be like) but that from axioms empty of useful knowledge and containing practically no information, like A=A, really complex and useful information can be derived by such simple axioms working in combination. This apparent difference I call Latent Information.

The game's afoot

The idea that from a few basic facts more information can be *deduced* is the key to Sherlock Holmes' success, but some caution is also required. The American Constitution it was hoped covered all eventualities of future society, but it didn't protect a basic right of people to freedom of worship, hence the addition of a further axiom, the First Amendment. Euclid would doubtless have said, "that's cheating!" Curiously, in England, the State *does* determines the people's worship: "the Church of England is, by Law established".

Computer programs are also axiomatic. You can't expect a computer to be cleverer than the program it is running.

Of course, you can add axioms or "patches" (or Amendments) as needs be and allow the system to expand and grow, but geometry is not like that. Euclidean geometry describes Euclidean space with a fixed set of independent axioms including the notorious fifth postulate ("parallel lines never meet"). Change that and you change the space to non-Euclidean hyperbolic and parabolic spaces.

In mathematics we seem to have a perfect body of Truth. That is the splendour of mathematics and its truthfulness has fascinated many.

Truth seeking and truth building

Two men fascinated by the power of mathematical truth and for much the same reason were Baruch Spinoza and Bertrand Russell. They both saw in geometry the possibility of an infallible even foolproof method of finding truth in general and not just in mathematics. Both men desired a way of truth building whose foundation was solidly and unshakeably certain.

And that is our task too.

Spinoza used the axiomatic method to try and prove a system of ethics. *Ethica Ordine Geometrico Demonstrata* (1677). Russell (with AN Whitehead) attempted to prove the whole of mathematics (including geometry) from an axiomatic system of symbolic logic. *Principia Mathematica* (3 volumes, 1910-13).

On being certain

Rene Descartes couldn't be sure of anything except his own existence; *cogito ergo sum*. His own existence he could not fail to doubt. (Mind you, Descartes never played the Glasgow Empire on a Saturday night!)

The beauty of geometric proofs is first its simple beauty. What impressed the young Bertrand Russell was that independent of anything in the world, $1+1=2$. Would that philosophy was like that. Would that you didn't have the uncertainties of religion and ethics.

So Bertie made a start. If you could prove everything in mathematics was a branch of symbolic logic, then perhaps eventually he could show that language itself could be made precise and the Truth about life and the Universe and everything could become as secure as $1+1=2$ (This, indeed, is proved as *54.43 in *Principia Mathematica to *56* (1964) p360)

$*54\cdot43.$ $\vdash :. \, \alpha, \beta \, \epsilon \, 1 . \supset : \alpha \cap \beta = \Lambda . \equiv . \alpha \cup \beta \, \epsilon \, 2$

    $Dem.$

      $\vdash . *54\cdot26 . \supset \vdash :. \, \alpha = \iota'x . \beta = \iota'y . \supset : \alpha \cup \beta \, \epsilon \, 2 . \equiv . x \neq y .$

      $[*51\cdot231]$ $\equiv . \iota'x \cap \iota'y = \Lambda .$

      $[*13\cdot12]$ $\equiv . \alpha \cap \beta = \Lambda$ (1)

      $\vdash . (1) . *11\cdot11\cdot35 . \supset$

        $\vdash :. (\exists x, y) . \alpha = \iota'x . \beta = \iota'y . \supset : \alpha \cup \beta \, \epsilon \, 2 . \equiv . \alpha \cap \beta = \Lambda$ (2)

      $\vdash . (2) . *11\cdot54 . *52\cdot1 . \supset \vdash . Prop$

   From this proposition it will follow, when arithmetical addition has been defined, that $1 + 1 = 2$.

Figure 1 – Principia Mathematica ✱ 54.43: From this proposition it will follow, that 1+1=2

As an empiricist, a stalwart follower of David Hume, young Bertrand would have been quite happy to throw out the baby (mentioned earlier) with the bath water. But then he hit a snag. The year was 1901.

How many Cretans can you get on the point of a pin?

The snag was irritating as he was later to tell Gottlob Frege who in turn thought there could be "hardly anything more unfortunate". The snag was simply "did the class of classes that didn't belong to itself, belong to itself or not? Don't even think about it. Instead think of the Islanders of Crete and imagine they all tell lies. But how do you know they are all liars? Because a Cretan said, "all Cretans are liars". But hang on a minute! If a Cretan said, "all Cretan are liars" and he's a Cretan, then all Cretans must tell the truth. But (and here we go) if he tells the truth when he says "all Cretan are liars" then all Cretans are liars, but he's a Cretan....and so it goes on.

It's that word "all" that causes the trouble. It includes itself. Bertie had a think about it one long summer and decided the best way was to cut it out like a cancer or gangrene. Don't mention "all" inclusively; but on what grounds? On the grounds that you'll go round in a vicious-circle if you do. Russell was quite happy to effectively ban certain types of statement, but when he mentioned the problem to Whitehead, Whitehead knew better. "Oh dear," he said, "never, glad, confident morning again!" (Quoting from Browning's poem *Lost Leader.)*

Of Towers of Babel & Golden Calves

Let's be clear about the *Principia* project. Russell was determined to build a Truth-system, man-made, complete, standing alone and absolutely true and consistent; in contrast to the **God** -made world of reality which to him appeared, messy, blurred, complex and largely incomprehensible. It was the stuff of which Golden Calves are built. And then came this snag. It was like a warning, "this could topple like a tower of cards" But he pressed on. With his *theory of types* he pressed on by simply excluding self-referential, circular statements, of the type "'all Cretans are liars,' (said a Cretan)". It was to prove to be an amazing mistake, with incredible consequences. With the original tower of Babel, **God** had to come right down to Earth to just to see this puny building Man was so proud of. The final volume of the *Principia* was completed in 1913 but the still small voice of protest came much later. The year was 1931 and the man responsible was Kurt by name and curt by nature.

Gödel's theorem

Gödel's theorem is straight-forward. In any axiomatic system (of reasonable size) you will always have, somewhere, at least one paradoxical statement, like "the Cretan who said, 'all Cretans are liars'", whose Truth or Falsity can not be **decided within that system**. Such a system as the Russell & Whitehead *Principia* will be either **incomplete** and **consistent** OR **complete** but **inconsistent** and never **complete** and **consistent**.

A Conjecture

Does this mean in effect that all systems which purport to explain everything; (theories of how society should behave; theories of what Man is, what is Nature is, what **God** is etc) are either **incomplete** and **consistent** OR **complete** but **inconsistent** and never **complete** and **consistent?** I will try and show it does.

I think why the *Principia* will always fail to be consistent or complete (even if you add additional axioms, the problem will still occur further on) is because it will always have a "loose end" an Omphalos. Like indeed we each have a "belly button" which shows us where we've come from. Man-made manufactured items have a trade mark, which doesn't belong to the item itself, but to the factory where it was made.

Russell & Whitehead (& Euclid) were part of the Universe they were describing. And not apart from it (insulated against real reality). This is a major error running all through Western thought beginning with the Greeks. We are part of the Universe.

Euclid's brain which could describe the whole of mathematics could also think up a new theorem which wasn't yet in the system. Why? Remember the Beatific Vision in which the mystic sees all joined up in One. Well perhaps they are! Therefore we can't isolate one item like a Tree and expect it to be complete in itself. Nothing is complete in itself, except the All. Another way of looking at the same thing; you can't make a copy of something in the real world without including "loose ends"; because everything in the real world has loose ends. Russell & Whitehead tried to make a copy of the world of mathematics, but mathematics is part of a larger Universe.

The Omphalos Conjecture

We suppose all things have "loose ends": (this is the generalised theory). It's only when we come to describe or explain (or copy) things *completely*, as did Gödel, that we realise we have to include "loose ends" even if they are paradoxical, because this is how they are in fact.

Gödel's proof[3]

Gödel never proved the Omphalos Conjecture for anything other than systems very like the *Principia*. This is because there is something very special about mathematics which just doesn't apply to, say, Trees. If this is taken into account it becomes a generalisation that if we *could* examine every item as minutely as Gödel meticulously examined the *Principia*, a similar conclusion could be made. The fact is we can't examine everything with Gödel's method but we do have the evidence of paradoxes. Hence the Conjecture is an explanation for why there are and must be certain otherwise inexplicable paradoxes in the world.

How Gödel did it

What Gödel did was construct a parallel or metamathematical system, carefully mapping and labelling each element he copied from the *Principia* in a analysable way. He used as labels to the axioms, prime numbers, in such a way as to make every theorem uniquely analysable. By factorising the number back into prime numbers each theorem could be traced back to the axioms on which it was based.

It would be rather like having a word for "Tree" which already contained words for "leaves" and "root" and words for "leaves" contained the whole biochemistry of photo-synthesis and so on, right down to the origin of trees. In this way it became clear to Gödel (and all who followed his argument) that the paradoxes came from an authentic and non-spurious source. They were part of life as we were meant to experience it.

Coping with the paradoxical

"Life is Life's way of coping with Life" or "Life is all about coping with Life"

The Möbius Twist

Figure 2- Möbius Strip

A Möbius strip made with a piece of paper and tape. If an ant were to crawl along the length of this strip, it would return to its starting point having traversed the entire length of the strip (on both sides of the original paper) without ever crossing an edge. (Wikipedia)

The paradox – through the Twist All becomes One
In the Möbius twist two sides of the paper strip become one – a continuous surface.

The twist is the Paradox

The Law of the Excluded Middle says "all statements are either True or False."

The paradoxical statement is; "are 'all statements are either True or False' True or False?" Now imagine on one side of the Möbius Strip every statement that is true and on the other all false statements. Join the Strip end to end with a Möbius Twist and this represents the paradoxical statement which is circular and self-referential.

The Strip is 2-dimensional and needs a third dimension to appreciate the significance of the Twist. The paradox is a "loose end" which moves us into another dimension.

Let's try another one. The Barber paradox. In a certain village all the men either shave themselves or are shaved by the Barber. Question: who shaved the Barber? Again the Barber, is reflexive and paradoxical. It indicates a "loose end". Not another dimension this time but a world larger than that of "men". A woman Barber would present no paradox. All the men could be shaved by themselves or by the Barber. No problem.

There are some recursive "loose ends". There is an a species of aphid apparently, which is born pregnant, by parthenogenesis, of course, but even so... All origin questions lead us to a circularity. Who made God? How did the Universe begin? What is the origin of life? What is human? Where did consciousness come from?

Will science remain incomplete?

So long as scientists themselves believe they are building a complete understanding of the world and its contents, science will remain incomplete. Science in any case, like everything else, is full of "loose ends". For instance, the very idea of "beginning" harkens back to the Judeo-Christian religious tradition and in particular to Genesis 1.1 "In the beginning God created the heavens and the Earth.."

References
1. see *Making use of Gödel's theorem*   Bulletin of the Society of Ordained   Scientists vol 34 (2004)
    The present analysis goes back to a PhD Thesis (Bradford, 1984) Appendix A.
    see also Proposition VI in *On Formally Undecideable Propositions in* Principia Mathematica *and Related Systems, I Monatschefte fur Physik* vol 38 (1931)
    pp   173-198 by Kurt Gödel.
2. *Omphalos* (literally navel i.e. tummy button)
    see Omphalos by P H Gosse (1857).
    The present analysis goes back to a MPhil Thesis (Hull, 1979).
    The *Omphalos Conjecture* is the generalised form of Gödel's theorem.
3.   see, for instance, E Nagel & JR Newman *Gödel's Proof* (1958)

Colin Price © 2013
Published in
Bulletin of the Society of Ordained Scientists,
Spring 2014

THE NATURE OF NATURE

CRAIG VENTER MEETS MOTHER CAREY ON A QUESTION OF LIFE

Craig Venter confidently predicts that within the next year he'll make new life on Earth by artificially engineering a dividing cell. This prediction is echoed by George Church who has already artificially constructed ribosomes, the protein assembly-lines of the cell.

From a News item

Taking the age of the Earth as about 4.5 billion years (byrs) then the first evidence of life has been estimated as occurring 3.8 byrs ago. Then the atmosphere had a large component of hydrogen which methanogens converted to methane. 2.3 byrs ago cyanobacteria began to rapidly fill the atmosphere with oxygen. As the methane content fell, the level of oxygen, poisonous to methanogens, rose. Both the methanogens and the cyanobacteria are autotrophs; the essential link between solar radiation and biological food i.e. sources of low entropy energy, upon which all subsequent life depends and makes up the food chain.

There is no problem about the energy balance for plant eating living things, it all comes ultimately from the Sun; the real problem is the origin of the autotrophs themselves. How did they get started and from where came the energy to enable them to take energy from the Sun? After all if it wasn't for photosynthesis, the Sun would merely heat up the planet and the heat would gradually radiate away. As it is the Earth's surface is now stacked deep, with low entropy, usable energy – vegetation, wood, oil, coal etc.

The observation which Charles Kingsley's Mother Carey made in *The Water-Babies* was "that it was not everyone who could make things make themselves". And this seems to me to underline the basic mystery of life. Not that she was the first to note it. When the famous philosopher Rene Descartes, tutor to Queen Christiana of Sweden, tried to explain to her that all living things are basically mechanical, and that even humans are merely mechanical devices with a material soul, the pineal gland, the grand lady, simply in refutation pointed to a clock and said "See to it that it produces offspring!".

In the 19th century a partial answer to the Darwin question was that "evolution was the way God made things make themselves" but with the absence of God in the present-day account this has rather left Charles Kingley's Mother Carey looking even more enigmatic. If God didn't first make things to make themselves, then who did?

If living forms are examples of things that make themselves then how do they do it? More significantly; is it possible, logically or otherwise for things to make themselves?

It is not clear if humans *can* make things that make themselves. This is what Craig Venter aims to prove one or another. But humans *don't* usually make things that make themselves. Car factories make cars but not car factories. During the war the Russians had factories inputting iron ore and coal at one end and a dozen miles later outputting tanks, not factories, at the other end. John von Neumann once envisaged in his *Theory of Self-Reproducing Automata* the human race one day making robot replicators, which would seek out raw materials and would eventually succeed unaided in covering the galaxy. Their sole purpose would be to multiply. This would be equivalent to making a dividing cell. But although they would be making things that make themselves, such robots would not have arisen by chance from nothing. Mankind has so far not made such a robot and no machine so far made has the compact ingenuity of living forms which, simply put, make oak trees from acorns. No machine has yet emulated the ingenuity of self-replicating macro-molecules. No machine builds like an organism; seemingly from the inside out, recursively. Organisms grow quite unlike how we humans would build things or would try to engineer life in a test tube. Organisms are holistic and markedly different to the rest of the material world. Nuclear synthesis, the making of the elements in the stars, by comparison, is conventionally understandable. Living things seem more complex than the interior structure of the stars!

Let us accept for the moment the unique peculiarity of living things.

It seems to me, thus, that the question *is it possible, logically or otherwise for things to make themselves?* has a paradoxical answer. No it's not and yes they do! Which has the same sort of bizarre logic as the Russell Paradox, *the class of classes which are not members of themselves is a member (of the class of classes which are not members of themselves)!* To try and pull the paradox apart slightly: it is said, all living things go back to the first dividing cell and the first replicating molecule which is how things can make themselves, but from where comes the animation which put together those first molecules and cells? Even von Neumann didn't suppose that the first robot replicator arose from nothing. It's the logical need of an external stimulus or input that brings about the paradox.

Simplified, the paradox comes down to this. *Yes it's living things that make themselves, but not life itself, that is simply passed on.* Logically, the process must have begun outside the present confines of living things; not here either on Earth or elsewhere in our galaxy but in another, but distinct, universe,. It's really the ship in the bottle, you need to be outside the system to initiate a life process of living things making themselves. That's the logic to Mother Carey's observation.

Craig Venter could further simplify the paradoxical observation of Mother Carey. He could say he is only a Cartesian mechanic, putting machine parts together. But even Mary Shelley's Dr Frankenstein anticipated there would be more to it than that. For this she added electricity. So the question, which Mother Carey would say, "No!" to, is this: Can the components of a cell, made artificially be put together to form a living cell that can make itself? Even then we may be led to ask, "how 'artificial' must artificial be?".

In the history of ideas, the *life process* straddles a huge spectrum, roughly from Descartes' view (as we have noted) as merely a part of the mechanical world to the mystical life force beyond the physical of Henri Bergson's *élan vital.*

I'm taking a median view. I envisage the life process as a pre-existing process in the history of the Universe. This is also found in the Biblical view. Note the distinct stages in Genesis 2.7 which also acknowledges the need of something else, external, that makes life the paradox it is. "And 1.God formed man from the dust of the earth and 2. breathed into his nostrils the breath of life and 3. man became a living soul." In the later Biblical tradition the figure of a pre-existing Wisdom makes a similar point. Something was there in the very beginning - *chokmah* (in Greek *Sophia)* described in the Wisdom literature (Proverbs 8.22ff).

Perhaps all life on Earth *can* be traced back to a single dividing cell, but that cell was not the *invention* of life. Indeed, the direct if somewhat strange analogy is with energy in a thermodynamic sense which is neither created nor destroyed. Energy too pre-exists the Universe. It was there before our Universe and will be there after our Universe disappears. Through time it spreads out and can do less and less. It is entropy which also gives a measure of disorder and in Lord Kelvin's words, indicates the "unavailability" of energy. This is what has been used up rather than what is available. Peter

Atkins doubts whether anyone really understands what energy is; entropy on the other hand he accepts as straightforward.

I don't profess to know what the "life process" is, except like energy it is eternal and both therefore pre-date the Big Bang. The bio-equivalent of entropy, I think, is more straight forward and comprises of the life forms and bio-systems the life process has played out, like the two epochs of the methane and oxygen bio-worlds mentioned in the opening paragraph.

We can only measure what has passed not what will come. What has been will never come again. It's as if life builds on (and needs) the debris of earlier epochs before it decides on what to do next.

Craig Venter & company may well prove, in substance, Mother Carey wrong and von Neumann right. We have just to wait and see. But to be a genuine *experimentum crucis* we need to know positively that the "new life" just made is not a new form of the old life.

In the 19th century there was a long debate on the nature of heat and energy before the concepts were clarified. We might be about to engage on the next phase in the long debate on the nature of life and its origin before we truly understand what the terms mean.

Colin Price © 2009

Why did God make Dinosaurs so
BIG?

It was little Amy who asked the question; "Why did God make dinosaurs so big?" All I could say was that not all dinosaurs were big - some were quite small - hen-size in fact.

Still, that was no real answer. I also mentioned whales being pretty big.

Whales!

All energy requirements for the living Earth come ultimately from only one source - the Sun and that through photosynthesis mainly in plants but also bacteria.

In the sea **phytoplankton,** step one in the food chain can grow and grow, as they depend on sunlight. Step two, is krill (crustacean) feed on phytoplankton. Step three, whales as predators of krill etc can eat and eat. Indeed what is there to stop whales becoming fewer in number and larger in size, as the krill multiply like mad? But hang on, by the same token why aren't the krill as big as whales? The quick answer is – they're not made to grow much bigger, but whales are. Oxygen is another factor, as we shall see.

Let's start again, Amy.

The first evidence of life on Earth occurred soon after the planet was formed 4.7 billion years ago, (byrs). The first life (probably **methanogens)** gave out methane and lived when the atmosphere was without any oxygen.

Then about 2.3 byrs ago and for a billion years oxygen levels steadily increased due to the bacteria **cyanobacteria** using photosynthesis and as a by-product releasing O_2 from water.

From them came the oxygen-using bacteria, fungi and animals and the CO_2 using plants. The two processes can be contrasted, the one used by plants, the other by animals.

photosynthesis - which uses $CO_2 + H_2O$ and yields low entropy energy (vegetation) + O_2. The vegetative material passes up the food chain.

respiration (food burned in O_2) - which releases energy + CO_2.

Note here the "food" is either from plant life or from animals which live on plants, herbivores.

So we seem to start with low energy phytoplankton and end up with .... high energy whales, and as we go up the food chain, the energy level increases. But this is an illusion. All the energy whales have to grow bigger and bigger still only comes from that one operation of photosynthesis which in turn depends solely on the energy from sunlight.

Now *photosynthesis* and *respiration*, oppose each other and today a balance has been struck which keeps the O_2 level at 21% of the atmosphere. This in effect is a balance between plants which photosynthesise and give off O_2 and multicellular animals which respire and burn up food, such as vegetation (or other animals) to obtain energy and give off CO_2 as a consequence.

For a billion years cyanobacteria kept converting CO_2 and filling the atmosphere with increasing amounts of O_2 – poisonous to methanogens and other bacteria who to this day survive as anaerobes, i.e. in the absence of oxygen.

So the first question is what was there to stop the plant life using up all the CO_2?

A good question. Fortunately most of the CO_2 is not in the atmosphere but captured as carbonaceous rocks from long dead animals.

During the Carboniferous period much carbon was buried becoming today, our coal deposits, effectively allowing the O_2 level to dramatically increase to 30% and allowing the insects to grow to enormous size, some as big as seagulls and millipedes a metre long.

This is certainly one way the animals in this case, insects, grew big.

Had this level of oxygen continued the Earth would have been set alight, for oxygen is highly inflammable.

So what stopped this happening ?

Perhaps I'm being naïve but I think the rise of respiring animals (and the sinking of their dead bodies as sedimentary rocks) saved the world from bursting into flames. In any case naïve or not I think the world of Nature was created by a thoughtful and wise Creator whose wondrously wrought creation fits together beautifully whether we believe it is a work of creation and fits together beautifully whether we don't believe it is a work of creation.

One can easily imagine a world whose landmass covers many continents, and adaptable dinosaurs and others faced with an over-production of goods roaming one vast world and simply growing and growing.

The pessimist Thomas Malthus calculated that the rise of the human population was outstripping food production and Charles Darwin assumed that this applied generally in Nature – in other words that there was a universal struggle for existence. This formed one of the key points of his theory of natural selection. But what was true for the human species which grew or caught its own specialised food was not necessarily true of Nature in general which was the assumption Darwin made. The vegetation is always increasing; but there will always be the possibility that Nature will find evermore ingenious ways of processing carbohydrates (from plants) and proteins (from animals) through a greater variety of species while another way is for some just to grow bigger.

Perhaps, Amy, the reason why God made dinosaurs so big was to enable them to eat all the accumulating vegetation and so to prevent the world from catching fire from a rising level of oxygen.

Colin Price © 2011

SWITCHING OFF THE MORAL COMPASS

A sense of morality may feel like a core part of our personality, but it can be easily switched off by sending a magnetic pulse through a region of the brain....

News item, Daily Telegraph, Mar 30 2010

And it doesn't just need a magnetic pulse, money, to name one non-magnetic source, can accomplish much the same result. It is not only the lust for money, but desires in general; what we used to call sin; wanton greed, vaulting ambition and things of that sort which cause dismay in the moral navigation department.

Indeed the more you think about it the more you come to ponder, "Well just what is it that keeps the old moral compass going?" Is there not a case for the Government here to step in and as it were to nationalise all moral compasses? Then we would have all our little compasses pointing in the same direction.

The Equality Bill presently undergoing parliamentary progress is perhaps a useful step in that direction. Under equality legislation, we will have little choice as to who we can invite into our homes, who make the better parents for adopted children, who ministers will marry in church, who the local church can employ and so on. No longer will there be excuses made for those of religious conviction, or "tender consciences".

So eventually consciences will all be under public ownership and we'll all be kitted out with identical items.

Meanwhile we have to face the fact that while some have no consciences at all, some of us obviously do. A case in point is the MPs' expenses row and the general public. The cry of; "they still don't get it!" shows that magnetic pulses are waving one way but not in others.

The present sad situation in the Catholic Church, whereby we have priests committing gross indignities on children over long periods of time, plus at the bishop level a covering up due to a misguided loyalty to the church, plus the Vatican complaining that they are being persecuted as if they were Jews facing anti-Semitism all goes to show that as was said about MP's expenses - "they just don't get it!" They too, can't see the immoral wood for the dodgy trees.

And what is it all these people don't get? Well, it's the consensual sum of moral compasses the rest of us have which allows us to distinguish right from wrong. Stand back and it's clear. Very clear. You ought to condemn all moral turpitude and if failure to do so would amount to perverting the course of justice you also have a civic duty to do so.

The Archbishop of Canterbury made things worse by retracting the claim he made that the Irish Church was losing credibility. Why? Because it is patently obvious that it *has* caused immense damage, both inside and outside the Church.

Where Rowan Williams is wrong is that it is not the Church that has failed, but certain clergy of the church. It is they who are failing their people – the catholic laity. It is they, the priesthood, not the laity who either committed or covered up or failed to warn or who ceased caring for the child in their midst – or let's face it, who continued to insist celibacy is a sacred virtue where too often among clergy it is a personal inadequacy.

The point that I don't have to make to Congregationalists is that the church is not the clergy. Would that the Irish

Catholic laity could call on a latter-day Martin Luther of their own to rid the church of corruption and start their own Catholic Reformation.

For it is clear to me that the Catholic church worldwide including the Vatican faces the biggest crisis of conscience since the Reformation. Certainly as Prof Eamon Duffy insists the French Revolution shook the church foundations, but that event was civic anarchy without, and not self-inflicted, moral anarchy within. At the Reformation, fingers pointed at the scandal of the Indulgences and the corruption within. Today, it is paedophilia and its shameful cover-up. Then, it needed a Counter-Reformation. Today, what?

Would that they had a Congregationalist to declare that the church is the people not its clergy, good or bad. It is its gathered church, (to use a Congregational term); the honest week-by-week worshippers. Let them rise up and claim the church for themselves and for Christians everywhere. But beware of magnetic pulses.

Colin Price © 2010

From the *Bulletin of Ordained Scientists*

No new Reformation - possibly a slow osmosis

Strange forces are sweeping through our English church life. I was talking to an American evangelist who was saying that she expected great things to emerge from the English churches.

When I protested and explained that our church attendance is down to 5%; our churches are nearly all struggling; and we are in the grip of a new secularism or, as it is sometimes called, militant atheism - and that's not even mentioning a rising Moslem population. Then when you look to the prosperous, well-filled pews of the Christian voice in America, how can you compare....

"Yes but," she replied, stopping me mid-sentence "you are at the cutting edge of change. You know what the problems are; you are hitting them head on, and you will be, through the grace of God, the first to find the solutions. After all you gave us the King James English Bible 400 years ago, next year, 2011."

"That was a long time ago," I feebly countered.

"It had along term effect," she retorted.

When I later mused on these ideas I began to think in terms of change, but what sort of change? There's no revolution, there's no new reformation, but there is an osmotic effect of diffusion through the traditional denominations.

You see it first of all in the Church of England, with issues over women (and homosexuals) in the episcopate. As I write the parish of St Peter in the diocese of Canterbury seeks Ordinariate status within the Church of Rome. It might be the first so to do but probably it will not be alone. Almost certainly they will not be taking church property with them; for unlike Congregationalists the diocese owns the church buildings, not the congregation; but the Church of England will be losing out when whole parishes seek to leave.

I was talking to Rev Ruth Whitehead, URC Minister, of Whitlesford in Cambridgeshire. She is also vicar of Whittlesford in an ecumenical partnership. The parish church and the chapel stand almost facing each other across the village green.

And is the division of Anglican and nonconformist still valid?

The URC, was set up in 1972 by Act of Parliament: how nonconformist is that? John Wesley founder of the Methodist movement lived and died an Anglican clergyman. And have you been to a Roman Catholic service lately? It's nothing like a High Church Anglican service. In fact it's more like a Low Church service, even, dare I say it, a free church service.

So, what is happening?

One: an unintended consequence of the New Secularism/Strident Atheism, is that it is galvanising once sleepy can't-be-bothered-going-to-church-any-more folk into the question; "Am I really an Atheist, then? - No I don't think so!"

Two - and this is really very important: people are slowly coming to the conclusion that their church is now WRONG, and they are going to do something about it.

Once it was URC members bemoaning the fact that they shouldn't have left Congregationalism. Now it is the righteous complaints of Catholic lay folk that their church was wrong to cover up the serious and criminal misdemeanours in the priesthood. And that the Church has got it fundamentally wrong over the unimportance of women. The Church's views on contraception are widely ignored, which for a centralised, dogmatic institution is not a point to be proud of. Lay folk can therefore now put forward the legitimate point that the work of Christ would be better carried out without the need for compulsory celibacy among the clergy.

The Church of England as we have seen is split irrevocably many ways. You just can't have one area served by two bishops and call yourself one church.

And Anglicans, in retrospect can legitimately say their church got it wrong over women priests. Either Synod should have made the issue a compulsory change of principle (like it presumably did in the early 19th century over the emancipation of slavery) – take it or leave it -- or it should have made no decision. Whatever the arguments once were, the fact remains that the church can now be seen to be in the wrong, either way. The Church has set up a system which made two ends of the church totally incompatible.

It is these senses of wrongness with the church that is powering strange forces for radical change.

If sufficient numbers, free church, Anglican and Roman Catholic are convinced their church has got it wrong, then change will inevitably follow.

I expect Catholic folk to quietly get on relying more and more on women lay leadership, keeping marriage behaviour strictly private and relying less and less on celibate, male, priestly leadership.

I expect the Church of England to quietly seep away from both ends: the evangelicals going the free church way and the Anglo-Catholics going to Rome.

I expect to see more Methodists (now covenanted with the Anglicans) and the URC (having lost half its membership since its inception) moving towards the middle ground of the Church of England and ecumenically serving our villages and small towns.

I expect to see new arrivals here, immigrants following Pentecostal ways and probably organised according to congregational principles of polity and revitalising Christianity in the face of an increasingly influential Islam.

Colin Price © 2010
Minister at Guilden Morden, Congregational Church, Cambridgeshire and Congregational Federation representative on the Theology & Unity Group of Churches Together in England. Written for *The Congregationalist*.

REVIEW

The Podfather –

the story of Robert Noyce and Silicon Valley,

A television programme broadcast on 12th October 2009
BBC4

There might not seem to be much of a connection between, say, Sir Titus Salt of Saltaire, Lord Leverhume of Port Sunlight fame and Dr Robert "Bob" Noyce of Silicon Valley, California but in the history of ideas, there are two common factors. All three displayed entrepreneurial innovation and not unconnected, all three had strong Congregational connections, which this television programme clearly brought out.

In this television biography of Bob Noyce, deemed the so-called "Mayor of Silicon Valley" his life was just as intimately involved in the place as was Sir Titus with Saltaire, save perhaps that, in an untypically American way, the Noyce name doesn't figure in the place name and in an untypically English way the name Salt does.

Robert Noyce came from a Congregational family. His father was a Congregational minister whose forebears went back to Governor William Bradford and of the first colony at Plymouth and William Brewster, both Pilgrims on the *Mayflower*. His mother's father, a Brewster, was also a Congregational minister. They lived in a self-sufficient way in a church dominated town in Iowa and the first college he attended, Grinnell College, was founded as Iowa College in 1846 by eleven Congregational ministers including the abolitionist minister, Josiah Bushnell Grinnell after whom the college was named in 1906 and to this day is renowned for its light-touch ethos of "self-governance" whereby much is allowed by the students and much is expected from them. The Congregational influence both in the man, the college and the place of Grinnell is well summed up in the phrase the journalist Horace Greeley once ascribed to Grinnell as saying, "go West, young man, go West, and grow up with the country!"

He was fortunate in his physics teacher, Grant O Gale, at Grinnell College with whom he learnt early on of the invention of the transistor by William Shockley, John Bardeen and others in 1947. He graduated in 1949 and afterwards took a PhD at Massachusetts Institute of Technology, MIT. In1956 he went to work for Shockley who was also instrumental in selecting the area in California which would become known as Silicon Valley. But Shockley, Nobel laureate, proved to be something of a dominating autocrat. In 1957 seven employees plus Noyce eventually broke away from the tight regime of Shockley Semiconductors and so started a pattern of change and innovation that made Silicon Valley a unique industrial complex that would in a few short decades dominate the globe. At Fairchild Semiconductors, founded by the "traitorous eight" Noyce co-invented (with Jack Kilby) the integrated circuit chip in 1960 which caused a revolution in the electronics industry.

Noyce's management style could be called "roll up your sleeves", which he (and others) ascribed to his Congregational upbringing. He shunned fancy corporate cars, reserved parking spaces, and private jets, in favour of a relaxed working environment in which everyone contributed and no one benefited from lavish perquisites. By declining the usual executive perks he stood as a model for future generations of Intel executives. He then founded with Gordon Moore, the company Intel. At Intel he oversaw Ted Hoff's invention of the microprocessor—that was Noyce's second revolutionary contribution to the electronics industry.

This then is the story of Silicon Valley and Bob Noyce. From the transistor came the printed circuit board, then the integrated chip, co-invented by Noyce, then the memory chip, and eventually the whole computer on a chip, Intel's microprocessor invented by Hoff.

Through all this Bob Noyce's equally innovative laid-back managerial method, whereby the whole company was a team whom he personally encouraged and the overall goal was simply new ideas, would become the engine room for the Valley's success. Among the names that Noyce directly mentored were Steve Jobs of Apple Computers and Bill Gates of Microsoft.

See also Leslie Berlin, *The Man Behind the Microchip: Robert Noyce and the Invention of Silicon Valley*, New York: Oxford University Press, 2005

Colin Price © 2009

From *The Congregationalist*

THE NATURE OF NATURE

REVIEW

𝔚𝔥𝔶 𝔓𝔯𝔦𝔢𝔰𝔱𝔰? - a failed tradition

By Garry Wills

Published by Viking NY (2013) $27.95

ISBN 978-0-670-0287-2

The author, a well-known American writer and Pulitzer Prize winner, a Roman Catholic layman, argues furiously what is wrong with his Church. In this case it's having a hierarchical structure embedded with priests whose *raison d'etre* is that they can perform the miracle of converting bread and wine into the body and blood of our Saviour, Jesus Christ.

These miracle workers once made (through the laying on of hands as in ordination by a bishop) can't have this miracle ability revoked, even if the priest himself is later "unfrocked". More curious this miracle worker doesn't need the presence of any further congregation to perform his (and never her) miracle. By himself it works, in Private Mass; in a mighty large congregation of fervent believers, but without a priest present it just doesn't.

There are further anomalies; the priest can't de-consecrate the elements, which leads to problems. "The un-de-consecrable Host, which is entirely and nothing else but the body and blood of Christ poses..." (p48) many difficulties which Thomas Aquinas himself considered; such as, if crumbs fall to the ground or if a fly accidently falls into the wine; but more strikingly one might worry, how does this holy mass pass through the intestinal system?

Some, like the Belgium theologian, Edward Schillebeeckx, pointed out there had to be a "reverse transubstantiation" to separate Jesus from the (Aquinian) "accidents" of bread and wine before they were excreted.

How did this absurd situation come about? I'm interested because as a Congregationalist I too believe in a truly lay church where if there is a priesthood, it is a "priesthood of all believers".

According to Wills, some of the first converts to the post-Resurrection church were priests from the Temple in Jerusalem who wanted to continue the sacrificial system. They presumably took the Lord's Supper which as the Last Supper was clearly a Passover meal and not much other than a remembrance; "do this in memory of Me" and made it into a sacrificial meal – the Mass. In order to counter this tendency the Epistle to the Hebrews was written, of which Wills provides a fresh translation. *Hebrews* explains that the only true priest was Christ the perfect priest (because he was sinless) making the perfect sacrifice (himself, freely offered unlike the pascal lamb in the earthly Temple) in the perfect place, Heaven. Because of this the Cross is once and for all time. There is only one priest, only one altar (Hebrews 13.1) and only one sacrifice, "once and for all". Or as Paul put it, "Christ *our* Passover lamb *has been* sacrificed". (1Cor 5.7)

The *Hebrews* argument presents clear reasons why the priesthood ought to be redundant, but, with most adjacent religions involving sacrificial rites, and some like Mithraism involving meal-like celebrations, Jewish sentiments gave way to the overpowering Roman influence. What was once a communal democracy, a "ministry of gifts" (*charisma*, 1Cor 12) was replaced by a "ministry of orders"; the hierarchical threefold ministry of deacon, priest and bishop. The Constantinian revolution put Christianity as the official religion. The title of *Pontifex Maximus*, head of the College of

Cults, first assumed by Julius Caesar was later claimed by the bishop of Rome, Damasus, from "the dust bin of history" having been disregarded by the emperor Gratian as being too pagan! So in effect the Roman Catholic Church soon became the new Roman Empire and the bishop of Rome, after the demise of the western wing in 410AD, the new Emperor.

Wills is right about priests (Greek *hierus* Hebrew *cohen*) and sacrifices. With no sacrifices after the Temple was destroyed in 70AD priests became redundant and to this day have remained so. The Cohen is still priest but plays little, even though an honoured, part in synagogue worship. His role disappeared with the end of the Temple and will only commence again with a re-constituted Temple which will surely herald the End Time!

Wills is wrong, of course, that the Lord's Supper, requires a miracle and that a priest is needed for that. Transubstantiation is a much later interpretation of the Eucharistic meal going way beyond the New Testament account of the Last Supper as Passover Meal and in any case it's no part of a priest's job to perform miracles. A Roman Catholic priest is an unnecessary intermediary between man and God. With the Incarnation the priest has no function in the post-Resurrection church. It is by the Incarnation that Jesus perfectly fulfils this role as Mediator, being both man and **God**, so that since the Risen Christ all believers can come to God through the intercession of the Son in the power of the Holy Spirit.

Meanwhile, why would the church adopt the title and function of priest? It is a good question. There is no mention of "priest" in the New Testament as a separate order except as Jewish priests or referring to Jesus as the High Priest. We, the Church, are a Royal Priesthood (1Peter 2.9) – all of us! We each have gifts which contributes to the whole. There is no one person more important than another. This was the ministry of gifts. Only much later did this change to the three-fold order of deacon, priest and bishop.

It seems to me, confusion started with a conflation of the following separate and quite distinct elements; the Last Supper, a meal; the Cross, as the means of redemption, the Pascal lamb, at home and in the Temple (until 70AD); the living presence of the risen Christ ("Peace be with you. As the father has sent me, so I am sending you. Receive the Holy Spirit" (John 20.21). The simple meal combined with the atoning nature of the Cross and sacrifice in the Temple all got laid upon one another.

Be that as it may, the Gospel message clearly separates Good Friday from Easter Sunday (day 1 of the Omer. Lev 23.15) and Easter Sunday from Pentecost (day 50 of the Omer), the coming of the Holy Spirit and the beginning of the Church - "whenever two or three are gathered together in His name." (Matt 19.20). There is no mention of churches being founded by bishops and led by priests. This came much later. There were certainly many external pressures to organise and systematise things, over the next centuries, as the Church grew, for instance, to make use of house-churches and to abandon the synagogue. Later, the *'church'* changed from being the people to being a building. The Gospel message was first given to both the apostles and the disciples through women in the garden and later many women like Priscilla, Lydia and Nympha hosted church assemblies, yet in spite of this the church increasingly denied an equal place for them. A similar thing was repeated as Methodism grew into a separate denomination. Two good innovations of John Wesley, women preachers and open air meetings, were soon abandoned as a more streamlined organisation developed.

All of this doesn't explain the peculiar position of the Anglican Church which doesn't accept the miracle of transubstantiation (see Article 28 of the 39 Articles in the Book of Common Prayer) yet still has priests. Why is this?

Priests (sacerdotalism) was the one of the main issues of the Reformation. (And before that, if you recollect the histories of the Cathars, the Hussites and in England, the Lollards.) Jean Calvin, made use of the New Testament word *presbuteros,* elder. In Presbyterianism the Elder is distinguished from the Minister and while both are ordained neither are priests.

But what about the Church of England which is Protestant but has priests?

Apparently   the   first   conversations   between   the Anglicans and the Methodist Church broke down when the Anglicans were told that the Methodist practice, after the Communion Service, was to throw the remaining bread out for the birds to eat. The Anglicans, although they didn't in their hearts believe in transubstantiation, behaved exactly like they would if they did. After a priest had prayed over the bread it couldn't still be bread and the wine, wine, could it? Yes it could! That's what we mean by not being a miracle. And it only becomes a symbol when it is swallowed as a symbol. The birds eat bread not symbols of the body of Jesus! Is that too hard to understand? Don't you see even by mentioning the word "priest" you muddy the waters. The word "priest" is just not needed.

The Jewish priest remains redundant – with no Temple there can be no sacrifices.

According to Wills, the RC priest *is* redundant if thinks he can perform the miracle of transubstantiation. In any case he is not a priest in the Jewish and Biblical sense – whose calling is to officiate sacrifices not miracles.

The Anglican priest ought to be redundant as he/she only has to preside over a memorial meal of the Lord's Supper where the elements are (only) bread and wine (Article 28). When you think about it there is nothing even vaguely "priestly" about consecrating (dedicating) the offertory (an offering) of bread and wine. As to the other functions of the Anglican priesthood, *benedictions* and *blessings* are given by God alone at the request of any good person and *confessions* are likewise best heard from spiritual friends not sacerdotal strangers.

Garry Wills does both the Roman Catholic Church and non-Episcopalian Protestants like the Congregationalists an invaluable service. Sadly, for neither and for quite differing reasons is his argument convincing. The Roman Catholic Church will continue to the End of Time with priests and bishops while Congregationalists will happily continue without them. But and, this is the point, he at least has explained his position, why no priests.

NOTE

from Article 28 of the 39 Articles in the Anglican Book of Common Prayer

.....Transubstantiation (the change of the substance of the bread and wine) in the Supper of the Lord cannot be proved from holy Scripture, but is repugnant to the plain teaching of Scripture. It overthrows the nature of a sacrament and has given rise to many superstitions. The body of Christ is given, taken and eaten in the Supper only in a heavenly and spiritual manner. The means by which the body of Christ is received and eaten in the Supper is faith. The sacrament of the Lord's Supper was not instituted by Christ to be reserved, carried about, lifted up or worshipped.

Intended!

Or

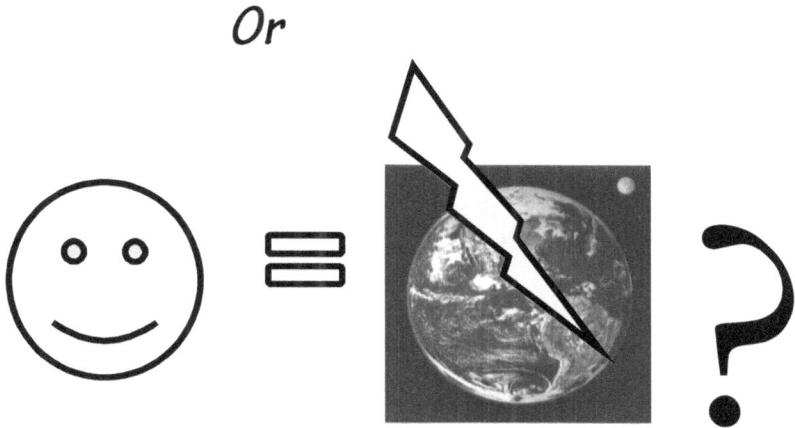

You don't have to be a fan of James Lovelock's Gaia theory to accept that the three great engines of change on Earth; the geo-physical system; the weather system and the bio-system are interrelated and it is human mentality and historical accident that divided them up. The three have this in common too; our theories of how they work are woefully inadequate as predictors of future events.

Contrast this to Newtonian physics, where gravitational theory very adequately and elegantly describes motion in the solar system. Was Newton lucky in his choice of problem? - or is there some intrinsic difference between the simplicity of the mechanical and the complexity of the geophysic/organic?

There certainly is. Mechanical systems are causal; containing the equivalent of levers and pulleys. Organic systems, whether living or non-living are holistic, are generally non-deterministic and non-predictable. To understand the difference between the two, consider a frog. You can prod a frog and it jumps "there"; prod it again in the same place and it jumps somewhere else. Organic systems are seldom simple and often have layers of redundancy; mechanical systems either work every time the same way or if they don't it's because they're broken!

In like manner there's probably no one simple way of describing Nature. Thus while John Ray marvelled at the gratuitous beauty of the butterfly; Tennyson viewed Nature, "red in touch and claw"; where Paley saw "design" Darwin saw, in Spencer's phrase, "the survival of the fittest" by natural selection. They all were trying to encompass the essence of Nature in one single thought or theory. It's as Oscar Wilde once put it "to every complex question there is a simple solution – and it's generally wrong". It's true of holistic systems.

Why seek Newtonian-type theories, that is, simple, single solutions to explain the whole range of complex phenomena that make up the natural world?

The brief answer is it is always worth a try in the hope you might again "do a Newton"

I suspect Darwin hoped he was "doing a Newton" for the natural world just as Freud hoped he was doing something similar for the conscious and unconscious mind. Karl Popper thought both were "pseudo-scientific" when held up to the test of falsifiability, but the difference was one not just of false and true science but of holistically and causally organized systems.

Darwin was honest enough to worry about alternatives like Paley's design. He could see natural selection answered a lot of questions but not all – some had to be shoe-horned into place. Then there are oddities like the duck-billed platypus, which are difficult to explain by any theory! Why ten sex chromosomes? (Humans have just two, X and Y!) Celestial mechanics never had these complications.

So what is an organic system like? Jesus seemed to put his finger on the nub when he says of the spirit to Nicodemus. *The wind blows where it wills....* As we have seen with frogs there is animation and autonomy in holistic systems. Even more impressive is Tyndale's translation, where "wind" is personalised and immediately evokes autonomy, mystery and will. *The wind bloweth where* he *listeth and thou heareth* his *sound but canst not tell whence* he *cometh or whither* he *goest.*

Compare this to a Newtonian framework, where "a body will continue in uniform motion until acted upon by a force..." This is predictable, causal, deterministic and pre- determinable. No question of one "canst not tell whence he cometh or whither he goeth." This body has its past and future mapped out. Such a body or a system made up of such bodies has no will or choice of purpose, no internal mystery. Such bodies act like billiard balls not frogs.

But are we talking of two different world views or two different systems in one world view?

Both really. Most fascinating still is the corporeal tendency of the parts of such systems to glue together a "whole" - almost in the way fractal images do the opposite, divide infinitely.

These parts join indefinitely. From eco-systems to plant and animal bodies, to the integration of cell activity to development of an embryo, there is a working together of the separate parts, even co-operative entities like insect and plant acting together in pollination.

Contrast this to causal systems, the essence of which is its analytic nature; they can be broken down and analysed; indeed this is the heart of the Cartesian method.

And here we come to the clock analogy showing the difference between causal and organic systems. Following the Cartesian method you can take a clock apart and work out the causal mechanism. But what does that tell you of the overall nature of a clock? In order to determine its purpose you have to treat it as a whole and ask what the overall function is. This is the limiting factor of analysis. When you take it apart it is broken and its function is lost. Holistic systems can't be conveniently analysed.

A further difference is the nature of "failed predictions". Causal predictions, that are valid can be verified and are potentially falsifiable *à la* Popper. In complex holistic systems things are not that straight forward. One remembers how Neptune was discovered and contrasts that to typical predictions in three major holistic systems; in the geo-physical system, the weather system and the natural world. Let's say predicting an earthquake; making an accurate weather forecast and predicting a next new species. With no great store of successes in these separate systems what confidence can be placed in those predictions across the systems. Can the natural world affect the climate? Yes. But how and in what ways is back to prodding the frog. Sometimes this causes that and sometimes this. As to the future who can say what will result from what. And what about, will reducing carbon emissions "improve" a climate change? I doubt it. Causal systems are often reversible, holistic systems more generally are not. Le Chatelier's principle of restoring an equilibrium which is what Gaia theory is all about, implies an irreversibility of cause and effect. You can't by prodding the frog again assume it will jump backwards.

This brings me to what I think is a major difference between causal and holistic systems. It is this: *in holistic systems all possible (but plausible) causal links end up being equally probable.* The frog will jump anywhere (possible) except backwards (implausible)

Why has the giraffe a long neck? When will this volcano erupt? What will the weather be like tomorrow? Even, when will this economic recession be over?

They all belong to holistic systems and we being "outside" we can't ever know exactly "whence, whatever it is, cometh and whither it goeth". They are like so many Just-So stories; they account for the facts without proving the facts. Contrast here Einstein's theory which accounted for the slow perihelion precession of Mercury as adduced on Newtonian theory and explained and accounted for the different predictions. Just-so stories are one among many equally possible accounts.

We can set up models of holistic systems. We know, for instance, that round the world the wind coming and going is, in total, constant. We have natural selection to account for certain biological events but without any powers of prediction. Weather forecasting, volcanoes, earthquakes, new species are all notoriously difficult to predict, just because they don't come out of causal systems.

Another way of putting this is to say that we can always trace any amount of plausible routes through holistic systems, most of which might be true but not all the time.

Sometimes, if only one such route is considered in the absence of more information, that one route is likened to a causal link in a causal system. Pavlovian reflexes in a dog might lead you think of dogs as causal, but they are organic and act holistically as we all know.

I must not give the impression that Newtonian, classical causal events are insignificant on the Earth. They are not. Physics and chemistry determine much of the events in geo-physics including the weather while organic chemistry and biochemistry determine much of the biosphere. But let's not forget that most of the atmosphere and large masses of the superficial geological structure of this planet are, in fact, biological detritus and without biological activity the Earth would be a very different place. And at the basis of which lies photosynthesis – so from where did that come? Which leads us back to the prior thought how and why did the first cell happen (of which we are all, from the first bacteria onwards its beneficiaries)?

The Earth hopefully is not unique, but the three holistic systems make it the strangest place in the Universe we yet know of. ***If it wasn't for the Earth, physicists would have a complete physical picture of the Universe from beginning to end.***

The barrier to a full understanding is the presence of that tiny speck and what is on it and what is in it. In the vast, vast Universe all can be conventionally understood except why the Earth is what it is. Yet without the Earth there would be ***no physicists doing physics and hence no mystery!*** Perhaps the ancients were right about this world having a special place in the Universe?

It is **INTENDED***!!*

Colin Price © 2011

www.ingramcontent.com/pod-product-compliance
Lightning Source LLC
Chambersburg PA
CBHW022026170525
45157CB00003B/1372